Geophysical Monograph Series

Including

Maurice Ewing Volumes

Mineral Physics Volumes

Geophysical Monograph Series

1 **Antarctica in the International Geophysical Year,** A. P. Crary, L. M. Gould, E. O. Hulburt, Hugh Odishaw, and Waldo E. Smith (Eds.)
2 **Geophysics and the IGY,** Hugh Odishaw and Stanley Ruttenberg (Eds.)
3 **Atmospheric Chemistry of Chlorine and Sulfur Compounds,** James P. Lodge, Jr. (Ed.)
4 **Contemporary Geodesy,** Charles A. Whitten and Kenneth H. Drummond (Eds.)
5 **Physics of Precipitation,** Helmut Weickmann (Ed.)
6 **The Crust of the Pacific Basin,** Gordon A. Macdonald and Hisashi Kuno (Eds.)
7 **Antarctic Research: The Matthew Fontaine Maury Memorial Symposium,** H. Wexler, M. J. Rubin, and J. E. Caskey, Jr. (Eds.)
8 **Terrestrial Heat Flow,** William H. K. Lee (Ed.)
9 **Gravity Anomalies: Unsurveyed Areas,** Hyman Orlin (Ed.)
10 **The Earth Beneath the Continents: A Volume of Geophysical Studies in Honor of Merle A. Tuve,** John S. Steinhart and T. Jefferson Smith (Eds.)
11 **Isotope Techniques in the Hydrologic Cycle,** Glenn E. Stout (Ed.)
12 **The Crust and Upper Mantle of the Pacific Area,** Leon Knopoff, Charles L. Drake, and Pembroke J. Hart (Eds.)
13 **The Earth's Crust and Upper Mantle,** Pembroke J. Hart (Ed.)
14 **The Structure and Physical Properties of the Earth's Crust,** John G. Heacock (Ed.)
15 **The Use of Artificial Satellites for Geodesy,** Soren W. Henriksen, Armando Mancini, and Bernard H. Chovitz (Eds.)
16 **Flow and Fracture of Rocks,** H. C. Heard, I. Y. Borg, N. L. Carter, and C. B. Raleigh (Eds.)
17 **Man-Made Lakes: Their Problems and Environmental Effects,** William C. Ackermann, Gilbert F. White, and E. B. Worthington (Eds.)
18 **The Upper Atmosphere in Motion: A Selection of Papers With Annotation,** C. O. Hines and Colleagues
19 **The Geophysics of the Pacific Ocean Basin and Its Margin: A Volume in Honor of George P. Woollard,** George H. Sutton, Murli H. Manghnani, and Ralph Moberly (Eds.)
20 **The Earth's Crust: Its Nature and Physical Properties,** John G. Heacock (Ed.)
21 **Quantitative Modeling of Magnetospheric Processes,** W. P. Olson (Ed.)
22 **Derivation, Meaning, and Use of Geomagnetic Indices,** P. N. Mayaud
23 **The Tectonic and Geologic Evolution of Southeast Asian Seas and Islands,** Dennis E. Hayes, (Ed.)
24 **Mechanical Behavior of Crustal Rocks: The Handin Volume,** N. L. Carter, M. Friedman, J. M. Logan, and D. W. Stearns (Eds.)
25 **Physics of Auroral Arc Formation,** S.-I. Akasofu and J. R. Kan (Eds.)
26 **Heterogeneous Atmospheric Chemistry,** David R. Schryer (Ed.)
27 **The Tectonic and Geologic Evolution of Southeast Asian Seas and Islands: Part 2,** Dennis E. Hayes, (Ed.)
28 **Magnetospheric Currents,** Thomas A. Potemra (Ed.)
29 **Climate Processes and Climate Sensitivity** (Maurice Ewing Volume 5), James E. Hansen and Taro Takahashi (Eds.)
30 **Magnetic Reconnection in Space and Laboratory Plasmas,** Edward W. Hones, Jr. (Ed.)
31 **Point Defects in Minerals** (Mineral Physics Volume 1), Robert N. Schock (Ed.)
32 **The Carbon Cycle and Atmospheric CO_2: Natural Variations Archean to Present,** E. T. Sundquist and W. S. Broecker (Eds.)

Maurice Ewing Volumes

1 **Island Arcs, Deep Sea Trenches, and Back-Arc Basins,** Manik Talwani and Walter C. Pitman III (Eds.)
2 **Deep Drilling Results in the Atlantic Ocean: Ocean Crust,** Manik Talwani, Christopher G. Harrison, and Dennis E. Hayes (Eds.)
3 **Deep Drilling Results in the Atlantic Ocean: Continental Margins and Paleoenvironment,** Manik Talwani, William Hay, and William B. F. Ryan (Eds.)
4 **Earthquake Prediction—An International Review,** David W. Simpson and Paul G. Richards (Eds.)

Geophysical Monograph 33

Greenland Ice Core: Geophysics, Geochemistry, and the Environment

C. C. Langway, Jr., H. Oeschger, and W. Dansgaard
Editors

American Geophysical Union
Washington, DC
1985

Greenland Ice Core:
Geophysics, Geochemistry, and the Environment

Library of Congress Cataloging in Publication Data

Main entry under title:

Greenland ice core.

(Geophysical monograph; 33)
1. Ice sheets—Greenland—Addresses, essays, lectures.
2. Greenland Ice Sheet Program. I. Langway, Chester C., 1929– . II. Oeschger, H. (Hans) III. Dansgaard, W. IV. American Geophysical Union. V. Series.

GB2596.5.G74 1985 551.3'1'09982 85-6015
ISBN 0-87590-057-7
ISSN: 0065-8448

Printed in the United States of America

Left to right: Duwayne M. Anderson, Associate Provost for Research, Texas A & M University (former Chief Scientist, Division of Polar Programs, U. S. National Science Foundation); Robert H. Rutford, President, University of Texas at Dallas (former Director, Division of Polar Programs, U. S. National Science Foundation); Willi Dansgaard, Professor, Geophysical Isotope Laboratory, University of Copenhagen, Denmark (Co-Convener of AGU/GISP Symposium); Chester C. Langway, Jr., Professor, Department of Geological Sciences, State University of New York at Buffalo (Co-Convener of AGU/GISP Symposium); Richard L. Cameron, Program Manager, Division of Polar Programs, U. S. National Science Foundation; James H. Zumberge, President, University of Southern California (former Member, National Science Board, U. S. National Science Foundation); Hans Oeschger, Professor, Physics Institute, University of Bern, Switzerland (Co-Convener of AGU/GISP Symposium).

This Volume is Dedicated to

HENRI BADER

Pioneering Ice Core Scientist

former Chief Scientist of the Snow, Ice and Permafrost Research Establishment,
U.S.A. Corps of Engineers, (USA SIPRE) Wilmette, Illinois

Support to C. C. Langway, Jr., provided by:
Coordinating Editor: Susan L. Kapuza, State University of New York at Buffalo
Copy Editor: H. Lorraine Oak, State University of New York at Buffalo
Proofs and Reference Lists Checked by: Dr. Hitoshi Shoji, State University of New York at Buffalo, and Dr. Jakob Schwander, State University of New York at Buffalo

CONTENTS

FOREWORD

The retrieval of cores from glaciers for scientific study was promoted during the planning phases of the International Geophysical Year (1957–58) by Dr. Henri Bader, who was then Chief Scientist of the Snow, Ice, and Permafrost Research Establishment (SIPRE) of the United States Corps of Engineers. Subsequently, the first deep ice cores were obtained from both Greenland and Antarctica.

The Greenland Ice Sheet Program (GISP) was the natural outgrowth of the interest and scientific return derived from the study of the ice cores from the surface of the Greenland Ice Sheet to its bedrock interface at Camp Century in 1966. This core was the object of significant research by scientists from Denmark, Switzerland, and the United States.

The need for additional coring was envisioned by this international group in order to extend the understanding of the history of the Greenland Ice Sheet. Accordingly, in the early 1970's, additional short cores were retrieved from central Greenland. Analyses of these cores made it clear that a more formal organization would be necessary to focus on the development of equipment, logistics, and techniques to accomplish the goal of obtaining a second core to bedrock from a Greenland site in a relatively simple glaciological regime. This effort was embodied in the GISP Science Plan of the mid-1970's. The GISP plan summarized the state of knowledge on ice core retrieval and analyses, outlined the scientific goals for future coring, and called for the development of the drilling, core handling, and design of equipment needed to accomplish these goals.

The GISP plan was approved and funded by the National Science Foundation and identified by that body as the major NSF scientific thrust in the Arctic during the late 1970's and early 1980's.

GISP represents a decade-long project that involved scientists from several nations. The results of their field work and laboratory studies which are reported in this volume have added a great deal to the interpretation of the history of the Greenland Ice Sheet and the climatic history of the earth. The development and application of the technologies and logistics proposed in the original GISP Science Plan made possible the achievement of most of the objectives by the scientists who designed the plan itself.

The GISP set a new standard for ice core retrieval and analysis, and will provide a basis for future ice core studies in Greenland, Antarctica, and elsewhere. To all who contributed to the GISP effort, the earth science community in general and the glaciological community in particular owe a debt of thanks and appreciation.

This AGU monograph represents the detailed work of the dedicated international team that conceived and implemented a complex logistical and sophisticated scientific program. Their accomplishments represent the best in modern glaciological research.

James H. Zumberge
University of Southern California
Robert H. Rutford
University of Texas at Dallas

PREFACE

The Greenland Ice Sheet Program (GISP) is already recognized as a major achievement in glaciology. GISP support came from the Swiss National Science Foundation, the Danish Commission for Scientific Research in Greenland and the United States National Science Foundation. And with the spirit, drive, and ability of Hans Oeschger, Willi Dansgaard and Chester Langway, GISP was planned, undertaken and successfully concluded. The results presented here demonstrate the significance of the climatic record stored in ice sheets and reemphasizes the need for additional deep ice cores from Greenland and Antarctica.

As you read these papers remember that these data were won with cold fingers, cold feet (both on the part of the field personnel and an occasional science administrator) and back-breaking work. Glaciological data is difficult to obtain, but the results are so rewarding that both the investigators themselves and the science community applaud the program.

As in many major scientific endeavors this too has been multi-national in scope with the principal countries being Denmark, Switzerland and the United States. Additional personnel involved came from Iceland and Japan. The melding of interests, enthusiasm, and scientific talent has enhanced the program. GISP did not always run smoothly but all involved knew the common objectives. The goal was reached together.

R. L. Cameron
National Science Foundation

THE GREENLAND ICE SHEET PROGRAM IN PERSPECTIVE

C. C. Langway, Jr.

Department of Geological Sciences, State University of New York at Buffalo, Amherst, New York

H. Oeschger

Physics Institute, University of Bern, Switzerland

W. Dansgaard

Geophysical Isotope Laboratory, University of Copenhagen, Denmark

Abstract. The Greenland Ice Sheet Program (GISP) was designed as a comprehensive interdisciplinary field and laboratory research effort to systematically determine the geophysical and geochemical characteristics of the ice sheet. The total program consisted of an array of separate but integrated science and engineering projects initially involving four principal research groups at: State University of New York at Buffalo (SUNY/Buffalo), University of Copenhagen, Denmark and the Technical University of Denmark, and University of Bern, Switzerland. Two of the major goals of GISP were mapping the thickness and the subsurface topography of the ice sheet, and continuous core drilling to bedrock. A radar survey was completed in 1974, and in the first drilling season at Dye 3 in 1971 and a 372-m deep, 10.2-cm diameter, ice core was recovered, followed thereafter by a nearly annual field operation: 1973 (Milcent, 398 m), 1974 (Crete, 405 m) and during 1979, 1980, and 1981 at Dye 3 where bedrock was reached at 2037-m depth on August 10, 1981 with a new deep drill.

The prime and broad objectives of the ice core investigations were to gain fuller understanding of the near-surface, internal, and subsurface nature, structure and composition of the ice sheet. Of special importance was the interaction between the dynamic and atmospheric processes affecting the ice sheet's formation, growth and decay, and its response to climatic change. Core studies incude stratigraphy, physical and mechanical properties, chemistry, microparticles, entrapped gases, stable isotopes, and radioactive isotopes. Airborne, surface, and borehole geophysical studies include radioecho soundings, magnetometry, gravimetry, strains and velocities, temperature and electromagnetic and sonic wave propagation. All major engineering and science goals of GISP were achieved.

Introduction

Core drilling into polar glaciers received modern impetus during the planning stages for the 1957-1958 International Geophysical Year (IGY) activities scheduled for both the northern and sourthern hemispheres [Bader, 1958; Crary et al., 1962]. Pre-IGY operations at Site 2 in N. W. Greenland during the summers of 1956 and 1957 [Lange, 1973; Langway, 1967] provided the experience required to develop the methods and techniques to successfully accomplish the ice core drilling objectives of the U. S. Antarctic IGY field programs at Byrd Station and Little America V, Antarctica [Patenaude et al., 1959; Ragle et al., 1960]. The overall success and widespread interest in the scientific results of the polar ice core drilling program were regarded significant enough to merit post-IGY continuation [Bader, 1962] at both Camp Century, Greenland [Ueda and Garfield, 1968] and Byrd Station, Antarctica [Ueda and Hansen, 1967; Ueda and Garfield, 1969].

The most recent initiative in polar ice core drilling to bedrock for scientific purposes started about 1970. Based on their past experiences, a group of engineers and scientists from the United States, Denmark, and Switzerland conceived and developed the Greenland Ice Sheet Program (GISP) as a joint cost-sharing and multidisciplinary study of the ice sheet. This program was designed to be accomplished by continuous core drilling at various locations with one of the ultimate goals being to core drill to bedrock. A comprehensive science, engineering, and operational plan was prepared and approved by the National Science Foundation's (NSF) National Science Board (NSB). [GISP, Science Plan, 1976].

The overall objectives of the Greenland Ice Sheet Program were to investigate the near-surface, interior, and subsurface of the ice sheet by continuous core drilling, core studies, surface reconnaissance, and remote sensing techniques. More specifically, GISP was designed as a three-dimensional investigation of the physical and chemical characteristics of the major physiographic, environmental, thermal, and dynamic zones of the Greenland Ice Sheet. The ice cores would provide the means to study the present and past mass balances, atmospheric processes, including climatic changes, and their impact on the ice sheet. The program consisted of an array of separate but integrated field and laboratory research projects developed and scheduled by the participating laboratories.

This symposium was planned 2 years before the field work was completed. Papers presented at the symposium emphasize the results of the various ice core studies, the new drill, airborne radar sounding, and ice flow modeling, with focus placed mainly on the Dye 3 ice core or studies made in the vicinity of the deep drill site.

Site Selection

The early plan for GISP included at least three deep core drillings at different locations on the ice sheet. However, as time went on, financial constraints made it clear that there would be only one deep drilling site within the framework of GISP. This limitation made it imperative that the drill site be chosen in the scientifically most favorable area to achieve the program objectives. Most of the field work prior to 1977 was aimed at finding an optimum site location. The basis of this search was the airborne radio-echo sounding that was performed in the period 1968-1974 and revealed the ice thickness and bedrock topography in most of the ice sheet area (Overgaard and Gudmandsen, 1978). The recovery and subsequent investigation of several cores to shallow (<100 m) and intermediate (<500 m) depths on the ice sheet (Figure 1, Table I) pointed to central Greenland as the most favorable area. However, since the drill had not yet been tested at great depths, the technically and logistically preferable Dye 3 location was eventually chosen as the operational site [Weertman et al., 1978].

Drilling

The original GISP field work began at Dye 3, Greenland in 1971 where, during a six-week field period, a continuous 372-m deep, 10.2-cm diameter ice core was recovered using the U. S. Army Cold Regions Research and Engineering Laboratory (USA CRREL) thermal coring drill [Rand, 1980]. This operation was followed by annual field forays to various geographical locations. During the early stages of GISP, two other intermediate-depth drilling operations were successfully accomplished by USA CRREL, one at Station Milcent in 1973 (398 m) and the other at Station Crete in 1974 (405 m).

The very first continuous drilling through a thick polar ice sheet to bedrock was achieved by USA CRREL in 1966 at Camp Century, Greenland [Ueda and Garfield, 1968]. Here 1385 m of somewhat continuous 10-cm diameter ice core was recovered. USA CRREL followed this success two years later by continuous core drilling 2164 m to bedrock at Byrd Station, Antarctica [Ueda and Garfield, 1969].

In 1970 the only ice coring drill proven capable of reaching bedrock was the USA CRREL electomechanical drill. However, it was determined at this time that a new device would have to be developed for GISP mainly because of the great bulk and weight of the USA CRREL drill which presented logistical problems. During the operational extent of GISP, improvements and modifications were made on several species of shallow and intermediate core drilling devices by the Universities of Bern, Copenhagen and by USA CRREL [see: Ice Core Drilling, J. Splettstoesser, Ed., 1976]. Development of the GISP deep ice core drill ultimately became the responsibility of the University of Copenhagen group which designed and constructed a completely new system. Some engineering support was obtained from both the University of Bern and the University of Nebraska-Lincoln. Drill development was a major requirement of the program and the resulting engineering milestone is the subject of a paper contained in this volume [Gundestrup and Johnsen, 1984]. Beginning with the summer of 1979, GISP started what turned out to be a three-season drilling-to-bedrock operation at Dye 3. The first year an 18-cm diameter casing hole was augered to 80 m [Rand, 1980] and extended to a depth of 225 m by core drilling with the new battery-powered and computer-controlled electromechanical ice coring rig. In 1980 ice core drilling penetrated to the 901-m depth and on August 10, 1981, bedrock was reached at 2037 m. The drill got stuck, but it was recovered in 1982. An excellent quality, 10.2-cm diameter ice core was recovered with less than 1 per mille of the core missing over the entire depth profile. With this achievement a major goal of GISP was accomplished, on schedule, and within the budgetary estimate as proposed in the science plan.

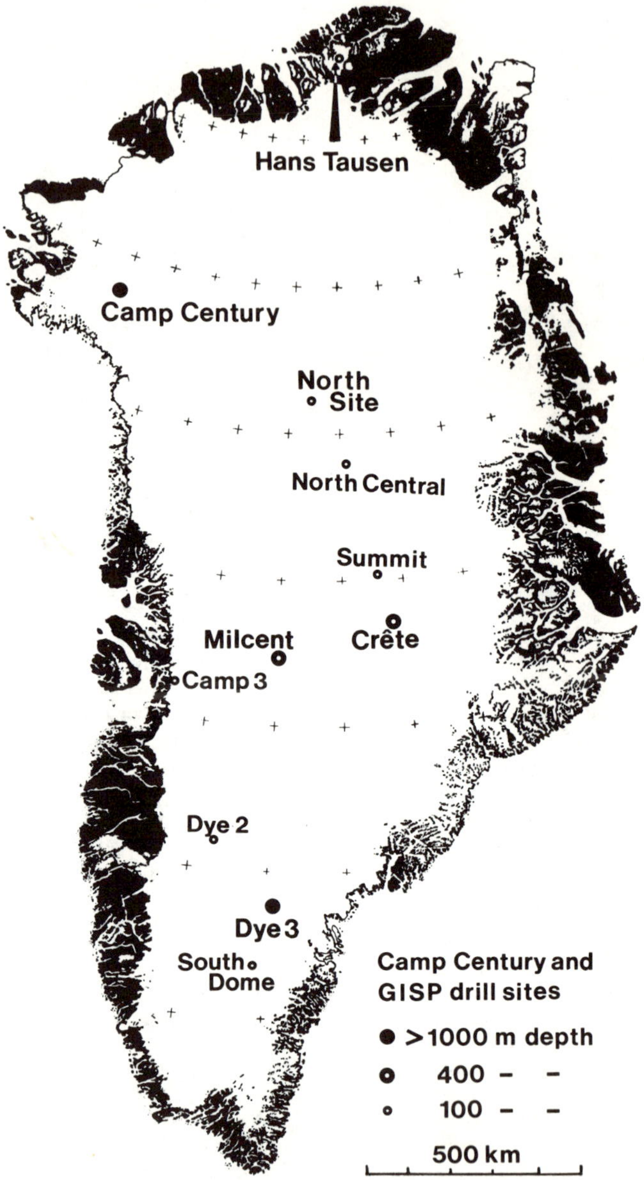

Fig. 1. Camp Century and GISP drill sites.

Core Studies

The vertical variations in the physical and chemical nature of the Dye 3 deep ice core is what reveals the paleoenvironmental record. The finite but elusive original layers are a product of discrete and variegated changes in the surface weather conditions at time of snow deposition. The layers represent definable limits measured in terms of specific events, seasons, decades, centuries or longer climatic intervals, but the physical layering becomes progressively more ill-defined and complicated with time (depth). These changes are caused by temperature fluctuations and physical alterations which occur as the deeper layers become subject to diagenetic processes and various stress conditions. On the other hand, except for very slow movement by diffusion, the isotopic and chemical substances

TABLE 1. GISP Drilling Operations

Year	Location	Coordinates	Type of Drill	Core Diam. (cm)	Depth (m)
1971	Dye 3	65°11′N, 43°49′W	Thermal (U.S)	10.2	372
1972	No. Site	75°46′N, 42°27′W	SIPRE (U.S.)	7.6	15
	Crete	71°07′N, 37°19′W	SIPRE (U.S.)	7.6	15
1973	Milcent	70°18′N, 45°35′W	Thermal (U.S.)	12.4	398
	Dye 2	66°23′N, 46°11′W	Shallow (Swiss)	7.6	50
1974	Crete	71°07′N, 37°19′W	Thermal (U.S.)	12.4	405
	Dye 2	66°23′N, 46°11′W	Shallow (U.S.)	10.2	101
	Summit	71°17′N, 37°56′W	Shallow (Swiss)	7.6	31
1975	Dye 3	65°11′N, 43°49′W	Shallow (Swiss)	7.6	95
	So. Dome	63°33′N, 44°36′W	Shallow (Swiss)	7.6	80
	Hans Tausen	82°30′N, 38°20′W	Shallow (Swiss)	7.6	60
1976	Dye 3	65°11′N, 43°49′W	Wireline (U.S.)	10.0	93
	Hans Tausen	82°30′N, 38°20′W	Shallow (Dane)	7.6	50
1977	Cp. Century	77°10′N, 61°08′W	Shallow (Dane)	7.6	100
	Dye 2	66°23′N, 46°11′W	Shallow (U.S.)	7.6	84
	No. Central	74°37′N, 39°36′W	Shallow (Dane)	7.6	100
	Camp III	69°43′N, 50°08′W	Shallow (Swiss)	7.6	49
1978	Dye 3	65°11′N, 43°49′W	Shallow (U.S.)	10.2	90
	Camp III	69°43′N, 50°08′W	Shallow (Swiss)	7.6	80
1979	Dye 3	65°11′N, 43°49′W	Thermal (U.S.)	(Access Hole)	0–80
			Electromech. (Dane)	10.2	80–225
1980	Dye 3	65°11′N, 43°49′W	Electromech. (Dane)	10.2	225–901
1981	Dye 3	65°11′N, 43°49′W	Electromech. (Dane)	10.2	901–2037 (Bedrock)

remain essentially as they were at time of incorporation. In general, the detailed environmental record becomes more difficult to interpret the deeper one proceeds, in particular when the Holocene to Pleistocene transition is passed, and in order to reconstruct a valid record, more reliance must be placed upon cross-correlations with other studies.

Most of the ice core studies were preplanned long before the deep drilling started. It was agreed upon by those involved that with the limited quantity of core material available and the great number of studies scheduled, that careful preplanning would ensure a maximum scientific return and offset the high costs of drill development and field operations. The careful planning assured that this new deep ice core would be the most completely recovered, thoroughly recorded, and comprehensively investigated ice core yet to reach bedrock.

We also feel that a new standard of efficiency was established for field science studies associated with the deep drilling. During a shallow-drilling operation, it is often desirable to minimize core handling in the field and simply make a stratigraphic log of the core. For an intermediate-drilling operation, limited studies are often made but are usually restricted to cutting specimens for specific studies to be performed later in the laboratory. The main factors determining the extent of studying the core in the field are the rapidity at which the core is recovered, the time scheduled for the drilling operation, and logistical considerations. A 100-m depth core usually can be recovered within 10 hours; a 400-m core within 4 to 6 weeks. A deep-drilling operation which requires several seasons is ideally suited for a more complete field study program, although this requires a larger operational base with extensive laboratory facilities and considerable personnel.

The field study techniques developed and adopted during GISP allowed for the entire ice core to be studied in continuous sequence on a line with stations for recoding logging, splitting, sample cutting and field analyses for acidity, dust content, physical and mechanical properties, and chemical composition. Special undersnow and surface laboratories were used (Fig. 2). This process provided immediate recovery, reduction and crosscheck of scientific data, and permitted the prompt preparation of several time-series of environmental parameters that previously took years to obtain during post-operational laboratory studies, all other factors being equal. The field procedures proved to be efficient and formative and allowed the first science results to be presented at the symposium less than one year after terminating the GISP field work.

Geophysics

A borehole to bedrock permits access to the inner nature of the ice sheet and provides various geophysical data at a point location (e.g., temperature, hole deformation, vertical strain, gravity, and electromagnetic wave velocity). However, additional measurements were needed over anisotropically larger areas for studying the dynamic behavior of the ice, e.g., bottom topography and rougness, horizontal strain, and velocity determinations along flow lines. The field methods used in the geophysical study program fell into two general categories: surface and airborne.

The surface traverses provided the means to obtain horizontal velocity data by use of portable geoceivers and navigation satellites [Mock, 1976; Drew et al., 1985]. Surface measurements of horizontal strain were made at rosette networks established at several local and regional sites [Mock, 1976; Whillans et al., 1984]. Airborne soundings were made using a C-130 type, ski-equipped aircraft with an inertial navigational system; mounted with under-wing antennae. Some detailed radio-echo soundings were also made during he surface traverses [Jezek et al., 1985; Drew et al., 1984] and subsequently in the borehole [Jezek and Roeloffs, 1985], but the majority of these data were obtained by airborne sounding [Gudmandsen, 1973, 1975; Overgaard and Gudmandsen, 1978].

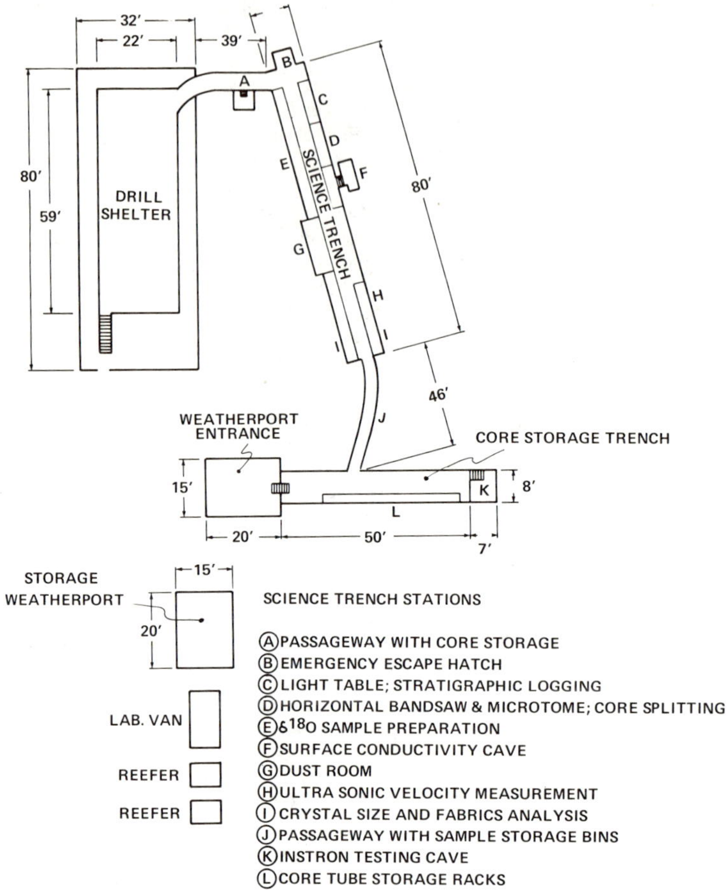

Fig. 2. Plan view of Dye 3 drilling and science trench. (Redrawn from Tillson and Kuivinen, 1982. Surface storage and entrance weather ports, laboratory and two refrigerators were positioned on the snow surface. The drill, science and storage trenches were located below the snow surface.)

Results

As was anticipated, the successful completion of the broad goals of GISP provided new scientific knowledge of the Greenland ice sheet and greatly supplement the earlier findings of the Camp Century and Byrd Station ice cores. The onsite and subsequent analyses of the Dye 3 ice core has substantially added to our understanding of the enviromental record, and in particular brought to light the abruptness with which major climatic changes have taken place.

From the Dye 3 results presented at the symposium we already have seen that:

1. Most of the pronounced oscillations observed in the oxygen-isotope profile from the deeper part of the Camp Century ice core were reproduced in the Dye 3 core, confirming their climatic significance and, at least, the regional validity of these data. A tentative link between continenatl and oceanic climatic records was also established. The Dye 3 core is being absolutely dated back to more than 7,000 years B.P. by seasonal cycles in ^{18}0, dust content, and acidity [Dansgaard et al., 1985];

2. Carbon dioxide measurements made on core samples show a considerable rapid increase in the atmospheric CO_2 concentration at the beginning of the Wisconsin-to-Holocene transition starting about 13,000 years B.P. [Stauffer et al., 1985];

3. The "acidity" (measured as solid state electrical conductivity) of the ice core was measured continuously and in great detail; the record reflects the Northern Hemisphere volcanic activity during the last 10,000 years [Hammer et al., 1985; Neftel et al., 1985];

4. Time-series analyses on the chemical composition of closely spaced ice core increments show a drastic and abrupt change occurs during the sharp Wisconsin-to-Holocene transition. The chemistry measurements will undoubtedly be a valuable tool for unraveling the cause of the remarkable change in the environment occurring at that time [M. Herron and Langway, 1985]; they also quantify the data obtained in the "acidity" study;

5. ^{10}Be concentration measurements made with the new accelerator mass spectrometry (AMS) revealed the solar modulation of radio-isotope production by cosmic radiaion (e.g., 11-year cycle, Maunder Minimum) and provide new information on changes in large-scale accumulation rates and/or cosmic ray flux [Beer et al., 1985];

6. Dust, ^{18}O, and acidity measurements show that the Bølling-Allerød-Younger Dryas oscillation was coincident in Greenland and Europe [Oeschger, 1985] and hence probably connected to rapidly

changing ice conditions in the North Atlantic Ocean;

7. Thin section and ultrasonic velocity studies provide detailed information on crystal size and orientation changes that are prerequisite input for realistic ice flow modeling [S. Herron et al., 1985];

8. New information was obtained on the mechanical properties of freshly drilled deep ice and other physical phenomenon such as air hydrates [Shoji and Langway, 1985];

9. Radar-sounding and strain-net measurements [Overgaard and Gundestrup, 1985; Jezek et al., 1985] outlined the regional upstream ice thickness, surface and bedrock topography, and ice velocities, all necessary parameters for ice flow modeling considerations;

10. Preliminary ice flow modeling results are promising, but the hilly bedrock upstream calls for further field work [Reeh et al., 1985; McInnes and Radok, 1985]. A perturbation approach flow model explains some of the apparent changes in the accumulation rate record as upstream effects [Reeh et al., 1985].

11. Other studies on artifical radioactive fallout [Koide and Goldberg, 1984, and lead concentrations [Patterson et al., 1984] demonstrate the broader potential of ice core investigations as sources of information on secular changes of the environment.

Post-Dye 3 Considerations

As with some symposia, the time spread has been long between the original presentation of the GISP papers and the resulting publication of these proceedings. Inevitably, additional results on GISP studies have been either presented at meetings or published in scientific journals [Dansgaard et al., 1984; Oeschger et al., 1984; Stauffer et al., 1984; Finkel and Langway, 1984]. In order to provide the reader with the current status of GISP, and to present our collective views on the anticipated future plans for core drilling in Greenland, the co-editors of this volume include this addendum.

The Dye 3 Site:

It will be a few years before the preplanned studies are completed on the Dye 3 ice core for all of the information it contains on the past environment; it will be even longer before all the data contained in the core is retrieved. Complicating the ice core analyses is the etremely complex ice flow trajectory over the hilly bedrock upstream of the recovery site. This required further field work in 1982 and 1983 in order to establish two of the fundamental time-series (i.e., annual snow accumulation and temperature change). Corrections in the measured raw data are dependent upon the upstream variations in the snow deposition rate and the geochemical composition, which were not fully known in 1981.

There are, however, other shortcomings related to the Dye 3 location. Due to present and past surface melting at the drill site during summer periods [M. Herron et al., 1981], spurious values are obtained for the CO_2 record. Atmospheric CO_2 is readily dissolved in meltwater and causes elevated CO_2 concentration levels in affected core samples. Regretably this situation invalidates some of the Holocene record (past 10,000 years) of the Dye 3 core as a source of valid information on one of the most urgent problems in modern climatology, i.e., the interaction between climate and atmospheric CO_2, including the turnover of CO_2 in the terrestrial reservoirs. Although we believe that no melting occurred during the pre-Holocene in the Dye 3 region, one cannot be sure. The CO_2 measurements reported at this symposium (Stauffer et al., 1984) and later ones, have attracted a great deal of attention at several recent international meetings [Keer, 1983; Campbell, 1984]. For complete confidence, this study must be validated on an ice core from a colder region

Finally, the Dye 3 core does not reach as far back in time as did the Camp Century core, at least not in continuous sequence [Dansgaard et al., 1984]. The mountainous bedrock relief upstream from Dye 3 has disturbed the stratigraphy in the deepest 87 m of the core. This makes it impossible to study a time period of great scientific interest: the Eem/Sangamon interglacial and its termination. Detailed studies of this event are important, because they might shed light on the significant problems of why, how, and when the present interglacial period will end.

A Central Greenland Site:

As mentioned earlier in the text, central Greenland was considered as early as 1977 to be the most appropriate location for deep ice core drilling. The rationale that was then developed was as follows:

1. The bedrock is relatively flat [Gudmandsen, 1973];
2. The ice divide is in very close proximity to the drill site;
3. No melting occurs on the surface;
4. and probably not at the ice/bedrock interface;
5. The interglacial accumulation rate is higher than 25 cm ice/yr;
6. and lower than 40 cm ice/yr;
7. The ice sheet is very thick (approximately 3100 m);
8. and the stable isotope profile correlates closely with climatic changes as far back as can be tested by systematic observations [Dansgaard et al., 1975].

The Dye 3 location only fulfills conditions numbers 4 and 6. Many of the difficulties being encountered with some interpretations of the Dye 3 ice core record will not be encountered in central Greenland.

The advantages of future ice core drilling in central Greenland are as follows:

a. Ice flow modeling will be simple, due to 1, 2, and 4, above (even though recent calculations show that the bottom temperature may be close to the pressure melting point).

b. Correction of a time series for upstream effects will be negligible or very small, due to 2.

c. The deposition sequence is most likely continuous as far back in time as the core reaches, due to 1, 2, and 5.

d. An ice core to bedrock will extend back in time two or three hundred thousands years, due to 4, 6, and 7, and hence reveal more than one glacial/interglacial cycle, if the bottom temperature had been below the pressure melting point that long.

e. Absolute dating will be possible back to at least 10,000 years before present with an accuracy that compares with dendrochronological result, due to 3, 5, and 7.

f. Detailed time-series of atmospheric CO_2 concentrations and other gaseous and chemical species will be measurable for glacial as well as interglacial conditions, due to 3.

g. The *in situ* annual layer thickness in 120,000-year old ice will be high enough to allow the last interglacial (Eem/Sangamon) and its termination to be studied in great detail, perhaps even on a year-to-year basis, due to 5, 6, and 7. This avails the prospect of determining the exact duration of the Eem/Sangamon termination, and the causes for it, e.g., if volcanic activity played an essential role.

h. The paleoenvironmental record revealed by chemistry and physical property measurements will be simple to determine and easier to interpret, clarifying and establishing significant index horizons.

i. The age-range of cosmic ray produced isotopes measured in the core will be compatible with the range of developed radioactive dating methods, namely the $^{14}C/^{12}C$, $^{81}Kr/Kr$ and $^{36}Cl/^{10}Be$ ratios.

j. The high acidity volcanic debris layers such as those found in the Dye 3 core at A.D. 1783 (Laki), A.D. 933 ± 2 (Eldgja), 50 ± 5 B.C. (unknown), and 8000 ± 100 B.C. (unknown) will be identifiable in another deep core, as well as from radio-echo sounding records in the central Greenland and the Jakobshavn basin area. It will be extremely valuable for ice flow modeling purposes to firmly establish these radio-echo isochrons.

Comparison of Greenland and Antarctic Sites:

Further drilling into both the Greenland and Antarctica ice sheets is considered to be of great need and broad significance for scientific reasons. Both of these continental-type ice sheets have some advantages and some disadvantages:

1. Studies to date show that the Antarctic ice sheet is acidic all the way to bedrock, which allows for general application of the simple and important space-charge method recently developed. [Hammer, 1980]. In contrast, Greenland ice that was deposited during the coldest part of the Wisconsin is alkaline, and volcanic activity in this period must therefore be measured by more laborious and extremely time-consuming analytical chemical methods.

2. According to preliminary calculations there is "only" 350 m of ice older than 10,000 years existing in central Greenland, compared to 2300 m at the South Pole, and 2850 m at Vostok Station, Antarctica. However, even 350 m of ice core is at least one order of magnitude more than is available from deep-ocean cores which, however, span the last few glacial cycles and consequently the central Greenland ice core will provide a finer degree of time resolution.

3. Antarctica, on the other hand, has some serious disadvantages: drill sites like Vostok and South Pole Stations do not fulfill conditions 1, 2, 5, nor perhaps 4 and 8 of the above described site criteria. In the Vostok Station area, for example, the present accumulation rate varies from 4 cm ice equivalent per year to negative values [Barkov et al., 1975]; this obviously means that the deep core from there is discontinuous in time. It has been reported that during interglacial periods, Vostok Station had a mean accumulation rate of 2.5 cm ice/yr, and only 1.0 cm ice/yr during glacial periods [Gordienko et al., 1982]. Furthermore, the South Pole Station is located hundreds of kilometers from the ice divide, and many of the difficulties experienced with the Dye 3 ice core are expected to occur in an enhanced form at the South Pole location. The mean accumulation rate at the South Pole under glacial conditions was probably of the same order as the present accumulation rate at Vostok (approximately 2.5 cm ice/yr), and therefore, the deposition rate is likely to be discontinuous. The favorable combination of site characteristics found in central Greenland: a simple flow pattern, and an acceptably high accumulation rate to ensure continuous deposition over an extensive time period, probably does not exist anywhere in Antarctica.

4. It is possible that a deep ice core from East Antarctica will extend further back in time than a core from central Greenland, but this is not an undisputed conclusion; it depends upon the ice/bedrock interface temperature, which is higher the lower the accumulation rate, when all other factors are equal. At Dye 3, the ice/bedrock temperature is −13C, and calculations show that it is somewhat higher in central Greenland.

Conclusions

Both the Antarctica and Greenland ice sheets are rich sources of information for Quaternary environmental conditions and processes acting upon the earth's surface during this period. A new long-term research program of core drilling to bedrock should systematically be initiated and carefully coordinated in both polar regions. This program should utilize the most advanced drilling and core study technology. Far-reaching scientific benefits would be realized with such an approach. Additional paleoenvironmental records from several geographically spaced deep ice cores would provide answers to the many questions of past climatic and environmental events, and better define, update, and contribute new proxy data for modeling projections on a hemispheric and global basis.

In short, we conclude that the deposition conditions are near perfect, and the ice flow is simple in the highest central part of the Greenland Ice Sheet. In fact, we believe that no other sedimentary sequence exists that offers as broad a spectrum of significant geophysical and geochemical parameters to measure long-term continuous time series. Our recommendation is that central Greenland should be given the highest priority for a future international deep ice core drilling program.

Acknowledgments. The total GISP program involved several institutes, laboratories, and universities. USA CRREL successfully core drilled at Dye 3 in 1971; Station Milcent in 1973; Station Crete in 1974; and bored the access casing hole at Dye 3 in 1979. Four principal research groups were initially involved in the science program: the State University of New York at Buffalo, U.S.A.; the University of Copenhagen, Denmark; the University of Bern, Switzerland; and the Technical University of Denmark. The University of Nebraska-Lincoln, U.S.A., managed logistical support activities and contributed to some engineering studies. During the decade of field activity, air transport and airborne radar sounding support was provided for by the U. S. Navy's VXE-6 Squadron, Pt. Hueneme, California; the U. S. Air Force National Guard's 109th TAG, Schenectady, New York; and the Royal Danish Airforce. The American units operated under contract arrangements handled by the University of Nebraska-Lincoln. Each university group was supported mainly by funding agencies within their respective countries. From its inception, GISP science was generously supported by the U. S. NSF's Division of Polar Programs (the Division also provided the major share of the operational and logistical budget); the Danish Commission for Scientific Research in Greenland; and the Swiss National Science Foundation. Later support was also received from the Carlsberg Foundation, Copenhagen, Denmark; and the Danish Natural Science Research Council. As GISP evolved, additional field and laboratory participants from Scripps, the University of Wisconsin, Ohio State University, and the University of Nebraska became engaged in other complementary geophysical, geochemical, and modeling aspects of the program.

The additional field and laboratory studies needed for the physical modeling of the flow characteristics upstream from Dye 3 were completed during 1982 and 1983 with funds obtained by the University of Copenhagen from the European Communities Climate Program (ECC), and U. S. logistical support sponsored by NSF.

References

Bader, H., United States polar ice and snow studies in the International Geophysical Year, American Geophysical Union, *Geophysical Monograph No. 2*, 177–181, 1958.

Bader, H., Scope, problems, and potential value of deep core drilling in ice sheets, *Special Report 58*, 6 pp., U. S. Army Corps of Engineers, Cold Regions Research and Engineering Laboratory, Hanover, N.H., 1962.

Barkov, N. I., F. G. Gordienko, E. S. Korotkevich, and V. M. Kotlyakov, Oxygen isotope studies on a 500 m ice core from a borehole at Vostok Station, *Information Bulletin Soviet Antarctic Expedition, 90*, 1975.

Beer, J., M. Andree, H. Oeschger, B. Stauffer, R. Balzer, G. Bonani, Ch. Stoller, M. Suter, W. Wolfli, and R. C. Finkel, ^{10}Be variations in polar ice cores, this volume, 1985.

Campbell, P., New data upset ice age theories, *Nature, 307* (5953), 688–689, 1984.

Crary, A. P., W. O. Field, and M. F. Meier, The United States glaciological researches during the International Geophysical Year, *Journal of Glaciology, 4* (31), 5–24, 1962.

Dansgaard, W., S. J. Johnsen, N. Reeh, N. Gundestrup, and H. B. Clausen, Climatic changes, Norsemen and modern man, *Nature, 255* (5503), 24–28, 1975.

Dansgaard, W., S. J. Johnsen, H. B. Clausen, D. Dahl-Jensen, N. Gundestrup, C. U. Hammer, and H. Oeschger, North Atlantic climatic oscillations revealed by deep Greenland ice cores, American Geophysical Union, *Geophysical Monograph 29, Climate Processes and Climate Sensitivity* (Maurice Ewing Series), 288–298, 1984.

Dansgaard, W., H. B. Clausen, N. Gundestrup, S. J. Johnsen, and C. Rygner, Dating and climatic interpretation of two deep Greenland ice cores, this volume, 1985.

Drew, A. R., and I. M. Whillans, Measurement of surface deformation of the Greenland Ice Sheet by satellite tracking, *Annals of Glaciology, 5*, 51–55, 1984.

Finkel, R. C., and C. C. Langway, Jr., Temporal variations in the chemical composition of ice cores as related to CO_2, ^{18}O, and dust records during the Wisconsin (abstract), Chapman Conference on Natural Variations in Carbon Dioxide and the Carbon Cycle, Tarpon Springs, Florida, January 9–13, 1984.

Gordienko, F. G., V. M. Kotlyakov, N. I. Barkov, and S. Korotkevich, Results of oxygen-isotope studies on the Vostok core, paper presented in the IAGP Scientific Session, House of Scientists, Leningrad, July 6, 1982.

Greenland Ice Sheet Program, Science Plan, Second Edition, 60 pp., Division of Polar Programs, National Science Foundation, Washington, D.C., October 1976.

Gudmandsen, P. E., Radioglaciology soundings at proposed drill sites, *D185*, 20 pp., Electromagnetics Institute, Technical University of Denmark, Lyngby, 1973.

Gudmandsen, P. E., Layer echoes in polar ice sheets, *Journal of Glaciology, 15* (73), 95–101, 1975.

Gundestrup, N. S., and S. J. Johnsen, A battery powered, instrumented deep ice core drill for liquid filled holes, this volume, 1985.

Hammer, C. U., Acidity of polar ice cores in relation to absolute dating, past volcanism, and radio-echoes, *Journal of Glaciology, 25* (93), 359–372, 1980.

Hammer, C. U., H. B. Clausen, W. Dansgaard, A. Neftel, P. Kristinsdottir, and E. Johnson, Continuous impurity analysis along the Dye 3 deep core, this volume, 1985.

Herron, M. M., and C. C. Langway, Jr., Chloride, nitrate, and sulfate in the Dye 3 and Camp Century, Greenland ice cores, this volume, 1985.

Herron, M. M., S. L. Herron, and C. C. Langway, Jr., Climatic signal of ice melt features in southern Greenland, *Nature, 293* (5831), 389–391, 1981.

Herron, S. L., C. C. Langway, Jr., and K. A. Brugger, Ultrasonic velocities and crystalline anisotropy in the ice core from Dye 3, Greenland, this volume, 1985.

Jezek, K. C., and E. A. Roeloffs, Measurements of radar wave speeds in polar glaciers using a down-hole radar target technique, *Cold Regions Research and Technology, 8*, 199–208, 1984.

Jezek, K. C., E. A. Roeloffs, and L. L. Greischar, A geophysical survey of subglacial geology around the deep-drilling site at Dye 3, Greenland, this volume, 1985.

Keer, R. A., The carbon cycle and climate warming, *Science, 222* (4628), 1107–1108, 1983.

Koide, M., and E. D. Goldberg, The historical record of artificial radioactive fallout from the atmosphere in polar glaciers, this volume, 1985.

Lange, G. R., Deep rotary core drilling in ice, *Technical Report 94*, 47 pp., U. S. Army Corps of Engineers, Cold Regions Research and Engineering Laboratory, Hanover, N.H., 1968.

Langway, C. C., Jr., Stratigraphic analysis of a deep ice core from Greenland, *Research Report 77*, 130 pp., U. S. Army Corps of Engineers, Cold Regions Research and Engineering Laboratory, Hanover, N.H., 1967 (also *Geological Society of America Special Paper 125*, 186 pp., 1970).

McInnes, B., and U. Radok, A steady-state prediction of Dye 3 core features, this volume, 1985.

Mock, S. J., Geodetic positions of borehole sites of the Greenland Ice Sheet Program, *Report 76-41*, U. S. Army Corps of Engineers, Cold Regions Research and Engineering Laboratory, Hanover, N.H., 1976.

Neftel, A., M. Andree, J. Schwander, B. Stauffer, and C. U. Hammer, Measurements of a kind of dc-conductivity on cores from Dye 3, this volume, 1985.

Oeschger, H., The contribution of ice core studies to the understanding of environmental processes, this volume, 1985.

Oeschger, H., J. Beer, U. Siegenthaler, B. Stauffer, W. Dansgaard, and C. C. Langway, Jr., Late glacial climate history from ice cores, American Geophysical Union, *Geophysical Monograph 29, Climate Processes and Climate Sensitivity* (Maurice Ewing Series), 299–306, 1984.

Overgaard, S., and P. E. Gudmandsen, Radioglaciology surface and bedrock contour maps at Dye 3, *R199*, Electromagnetics Institute, Technical University of Denmark, Lyngby, 1978.

Overgaard, S., and N. S. Gundestrup, Bedrock topography of the Greenland Ice Sheet in the Dye 3 area, this volume, 1985.

Patenaude, R. W., E. W. Marshall, and A. J. Gow, Deep core drilling in ice, Byrd Station, Antarctica, *Technical Report 60*, 7 pp., U. S. Army Corps of Engineers, Snow, Ice, and Permafrost Research Establishment, Wilmette, Illinois, 1959.

Patterson, C. C., C. Boutron, and R. Flegal, Present status and future of lead studies in polar snow, this volume, 1985.

Ragle, R. H., B. L. Hansen, A. J. Gow, and R. W. Patenaude, Deep core drilling in the Ross Ice Shelf, Little America V, Antarctica, *Technical Report 70, Parts I and II*, 10 pp., U. S. Army Corps of Engineers, Snow, Ice, and Permafrost Research Establishment, Wilmette, Illinois, 1960.

Rand, J., 1979 Greenland Ice Sheet Program, Phase 1: Casing operation, *Special Report 80-24*, 18 pp., U. S. Army Corps of Engineers, Cold Regions Research and Engineering Laboratory, Hanover, N.H., 1980.

Reeh, N., S. J. Johnsen, and D. Dahl-Jensen, Dating the Dye 3 deep ice core by flow model calculations, this volume, 1985.

Shoji, H., and C. C. Langway, Jr., Mechanical properties of fresh ice core from Dye 3, Greenland, this volume, 1985.

Splettstoesser, J. F. (Editor), *Ice-Core Drilling*, 189 pp., University of Nebraska Press, Lincoln, Nebraska, 1976.

Stauffer, B., U. Siegenthaler, and H. Oeschger, Atmospheric CO_2 concentrations during the last 50,000 y from ice core analysis (abstract), Chapman Conference on Natural Variations in Carbon Dioxide and the Carbon Cycle, Tarpon Springs, Florida, January 9–13, 1984.

Stauffer, B., A. Neftel, H. Oeschger, and J. Schwander, CO_2 concentration in air extracted from Greenland ice samples, this volume, 1985.

Tillson, R. A., and K. C. Kuivinen, An ice core science trench for use by glaciologists on the Greenland ice sheet, in *Antarctic Logistics Symposium Volume*, 7 pp., Scientific Committee on Antarctic Research, Leningrad, 1982.

Ueda, H. T., and B. L. Hansen, Installation of deep core drilling equipment at Byrd Station (1966–1967), *Antarctic Journal of the U. S., 2* (4), 120–121, 1967.

Ueda, H. T., and D. E. Garfield, Drilling through the Greenland ice sheet, *Special Report 126*, 7 pp., U. S. Army Corps of Engineers, Cold Regions Research and Engineering Laboratory, Hanover, N.H., 1968.

Ueda, H. T., and D. E. Garfield, Core drilling through the Antarctic Ice Sheet, *Technical Report 231*, 17 pp., U. S. Army Corps of Engineers, Cold Regions Research and Engineering Laboratory, Hanover, N.H., 1969.

Weertman, H. (Editor), Considerations for the selection of the Greenland Ice Sheet Program deep ice core drilling site, 7 p., U. S. GISP Science Committee, Division of Polar Programs, National Science Foundation, Washington, D.C., March 1978.

Whillans, I. M., K. C. Jezek, A. R. Drew, and N. S. Gundestrup, Ice flow leading to the deep core hole at Dye 3, Greenland, *Annals of Glaciology*, *5*, 185–190, 1984.

THE CONTRIBUTION OF ICE CORE STUDIES TO THE UNDERSTANDING OF ENVIRONMENTAL PROCESSES

H. Oeschger

Physics Institute, University of Bern, Switzerland

Abstract. Data obtained from the studies of polar ice cores supplement the records available from tree rings, peat bogs, lake and ocean sediments, and provide a relatively new data source to understand processes of the complex climatic and global cycles. The main sources of ice core data are stable and radioactive isotopes, soluble and particulate matter, and the composition of the gases occluded in the ice. Such information can be used to investigate the history and the variability of carbon dioxide and the climate system.

Temperature and other climatic data obtained from $\delta^{18}O$ measurements of polar ice cores can be correlated with similar information obtained from carbonate lake sediments. Comparison of the $\delta^{18}O$ profiles of the Dye 3 ice core and central European lake sediments show distinct similarities such as the identification of the Older Dryas–Bølling/Allerød–Younger Dryas–Preboreal sequence.

Measurements of the cosmic ray produced isotope ^{10}Be on only 1 kg polar ice samples are possible by accelerator mass spectrometry. The resulting data reveals the 11-year solar modulation cycle and the Maunder Minimum of solar activity from 1645 to 1745 AD. The ^{10}Be concentration values for the Maunder Minimum are a factor 1.6 higher than the average for the past 800 years. Using a carbon cycle model these ^{10}Be variations can be compared to the ^{14}C variations found in tree rings. The relatively good correlation suggests a common origin of the ^{10}Be and ^{14}C fluctuations and serves as a check of carbon cycle models. During the Wisconsin stage all of the Dye 3 ice core parameters measured to date ($\delta^{18}O$, CO_2/air, $SO_4^{=}$, NO_3^-, Cl^-, dust) show values fluctuating between two different boundary conditions. This suggests that the climate system existing at that time oscillated between a cold and a warm state, probably strongly influenced by different ocean circulations and ice cover. During the Wisconsin stage a cold system dominated; the transition to the Holocene is considered as the final transition to a warm state. Thereafter the boundary conditions did not allow the systems to switch back to a cold state.

Introduction

To improve our understanding of the complex processes that determine our environment is an old and fundamental scientific goal. However, in today's complicated society, many new and significant questions are raised in this traditional area of scientific inquiry, because man's impact on the global environment appears to produce serious increases in various environmental substances above and beyond their natural fluctuations. For example, the sustained emission of fossil fuel CO_2 contributes to higher atmospheric CO_2 concentrations that might very well induce a rapid climatic change. To assess the consequences of anthropogenic CO_2 emissions we require additional information on the carbon cycle to estimate how much of the released CO_2 will remain in the atmosphere, and about the climatic mechanisms involved, to obtain an idea of the climatic impact of the increase of CO_2 concentrations that are expected.

During the past few years significant progress has been made in understanding certain environmental processes, and major contributions can be expected within the next decade mainly for the following reasons:

1) The methods of observing present earth-surface environmental conditions have been greatly improved and coordinated on an international global network basis supported by satellite data.
2) New samples of earth materials, such as ice, organic matter, and sediments have become available. These samples have been carefully archived, and the results of various studies have enabled scientists to reconstruct, in some cases in great detail, environmental changes in the past. In this context the ice core drilling through the Greenland ice sheet at Dye 3, South Greenland, deserves special attention.
3) Chemical and physical analytical techniques have been newly developed or perfected which enable rapid, reliable, continuous or quasi-continuous measurements of natural parameters. One example is the accelerator mass spectrometer (AMS) [Muller, 1977; Bennett et al., 1977; Suter et al., 1981] which, due to its extremely high sensitivity, enables measurements of certain cosmic ray produced radioisotopes at levels of only 10^3 to 10^6 atoms.
4) Progress in computer technology offers an increasingly realistic simulation of the complex processes taking place in natural systems.

In this paper we first introduce the concept of environmental system research, emphasizing especially the value of the information on past system states recorded in polar ice cores. For illustration we use examples related to the CO_2 system and the climate system research. Finally we summarize some of the most important recent results obtained mainly on the Dye 3 ice core and briefly discuss their scientific relevance.

The Environmental System (E.S.)

To illustrate our research approach we introduce the concept of a global environmental system [Oeschger, 1982]. It is illustrated in Figure 1. It includes the entirety of physical, chemical, and biological processes acting upon the earth's surface and in the atmosphere. The various parts of the system interact in various dynamic sequences. The main components of the E.S. are the *atmosphere,* divided into the troposphere and stratosphere; the *hydrosphere,* including the oceans and continental waters; the *cryosphere,* consisting of the polar ice sheets, sea ice, and mountain glaciers; the

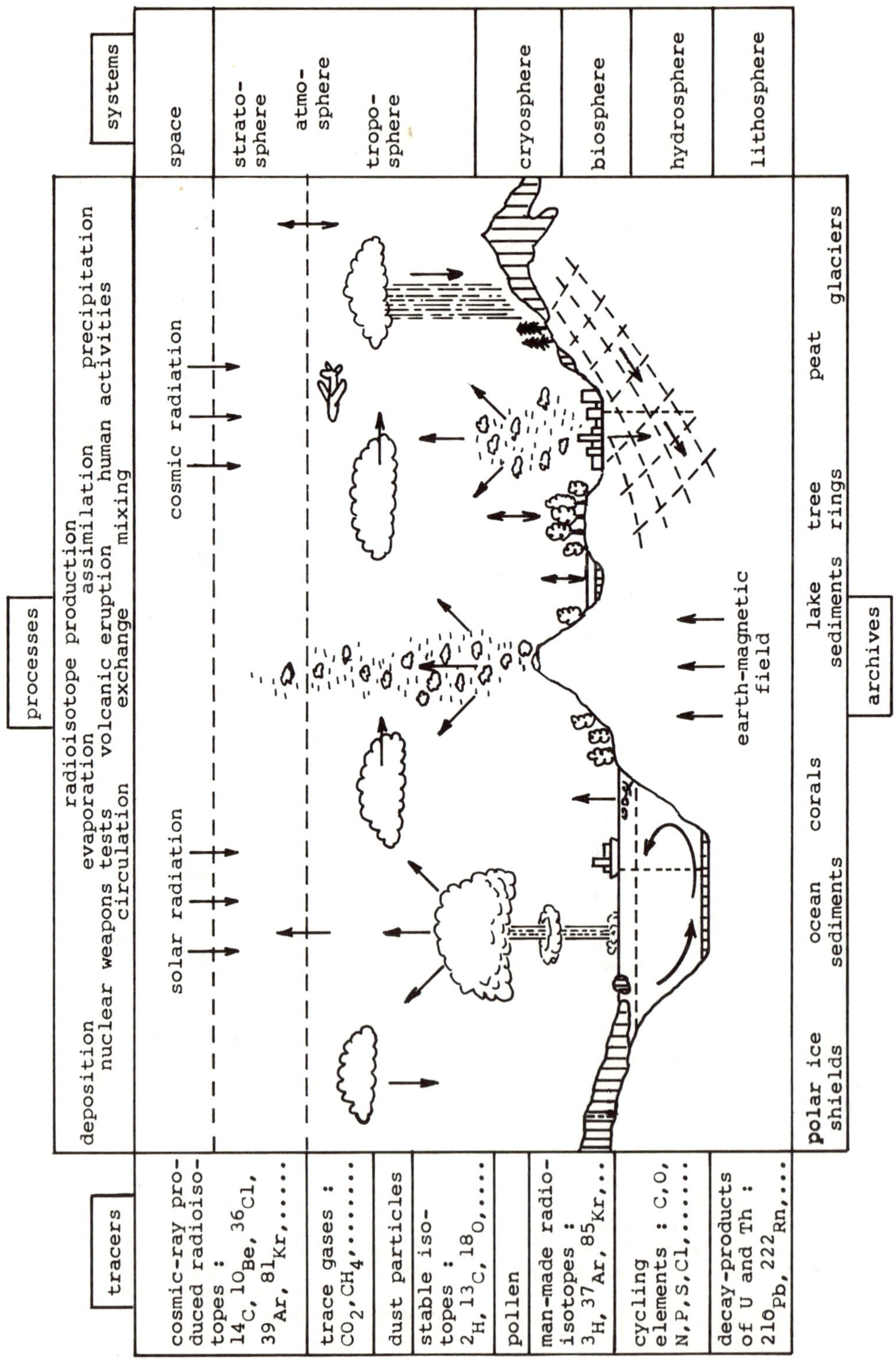

Fig. 1. Diagram of the environmental system concept. (Designed by Drs. B. Lehmann and J. Beer, Bern)

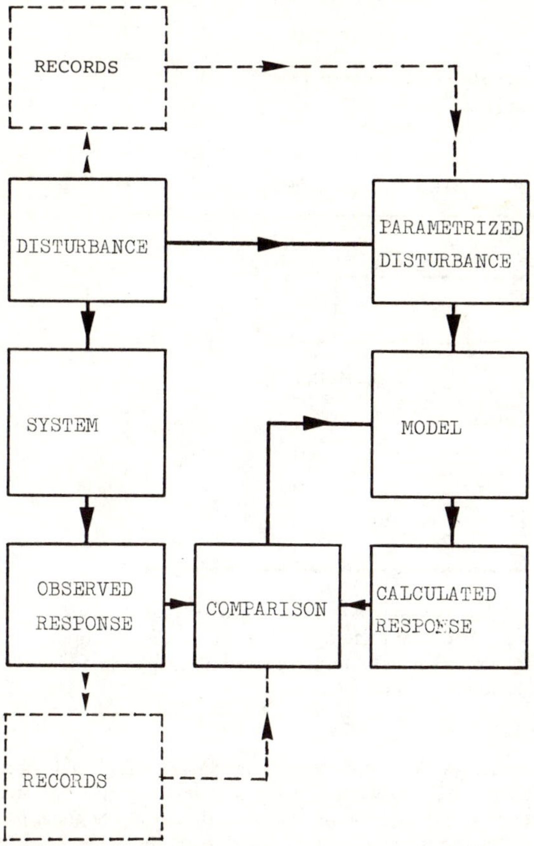

Fig. 2. Flow diagram illustrating the approach to understand environmental processes. The left column represents the actual environment and its processes, the right one the model simulation which is also used for prognoses based on assumed disturbance scenarios.

biosphere, containing the marine and continental living organisms; and the *lithosphere,* with bedrock and sediments that interface with the hydrosphere.

The sun's energy drives the dynamic processes in the E.S. It causes atmospheric circulation and oceanic mixing and due to evaporation and precipitation, the cycling of water. Control of the energy balance, which in turn determines the climatic conditions at individual locations on the earth's surface, is affected by scattering and reflection of short-wavelength solar radiation in the atmosphere and on the earth's surface, as well as infrared radiation and emission absorption and reemission by the surface and by water vapor and gases (CO_2, O_3, . . .) in the atmosphere, together with the transfer of latent and sensible heat into the atmosphere and ocean. The dynamic cycles of some elements, such as C, N, and O, play an important role that is regulated by biospheric activity, chemical reactions, and physical exchange processes.

Dust particles and aerosols are injected into the atmosphere by wind action, volcanic eruptions, biospheric processes, and human activities. They reappear on the earth's surface by dry fallout or wet deposition. Of special interest for our approach to the understanding of the E.S. processes are the radioactive and stable isotopes that constitute ideal tracers for a variety of processes and their dynamics.

Radioactive isotopes have different origins:

a) They form by interaction of cosmic radiation with atoms and molecules existing in the upper atmosphere (^{10}Be, ^{14}C, ^{36}Cl, ^{39}Ar, ^{81}Kr, . . .). This cosmic production is modulated by changing shields of solar plasma and the earth's magnetic field.
b) They are also introduced to the E.S. as a result of man's use of nuclear fusion and fission (^{3}H, ^{37}Ar, ^{85}Kr, . . .); and finally,
c) They are released from the earth's crust as products of the natural decay series of U and Th (^{222}Rn, ^{210}Pb, . . .).

If the initial concentration of a radioactive isotope in a certain sample after its separation, e.g. from the atmospheric or oceanic reservoirs, is known, the actual concentration can be used to calculate the age of the sample, i.e. the time elapsed since the separation. Since halflives vary from days (^{37}Ar, ^{133}Xe) to hundreds of thousands of years (^{10}Be, ^{36}Cl, ^{81}Kr), information is attainable about time constants of natural processes over a very wide range. Each isotope has its characteristic field of applications which may reach far beyond that of merely dating samples. Stable isotopes (^{2}H, ^{13}C, ^{18}O) are another important source of information. Phase transitions, chemical reactions, and diffusion processes produce small changes in the natural isotope ratios; they reflect such things as the temperature at which a phase transition occurred.

Samples of air, water and ice, organic materials, and sediments, taken from any part of the E.S., contain information on its *static characteristics,* like the partitioning of water between atmosphere, biosphere, and ocean, but also its *dynamic characteristics,* such as mixing and circulation and exchange processes in and between the different system components. This and other information can be derived from the isotopic ratios (stable and radioactive) and from the concentration levels of chemical elements and molecules, pollen, and dust. A complete set of these parameters defines the state of the E.S. These "fingerprint parameters" are continuously being recorded in the natural archives or reservoirs of the E.S., such as polar ice sheets and mountain glaciers as well as ocean and lake sediments and organic materials (tree rings, peat, and coral deposits). These earth reservoirs, thus, constitute the memory bank of the E.S. and the analyses of sequential samples over temporal profiles allows the reconstruction of its historical evolution.

Environmental System Research

The overall objective of investigating the E.S. is to obtain a consistent and systematic description of its complex processes and an understanding of the mechanisms by which they are controlled. Mathematical models play a large role in E.S. research. Models are based on the fundamental physical, chemical, and biological knowledge. But the various processes are very complex. It is often necessary to simplify the equations and to use empirical procedures to arrive at conclusions. The models are validated by testing their capability to produce system responses to various perturbations that closely agree with observation. Man's impact upon the environment is an example of a perturbation. Past system perturbations and responses are revealed from the natural archives. Figure 2 illustrates the process of model development. Such an analysis leads to increasing confidence in the models which also are used to predict future natural or anthropogenic environmental changes.

During the past years the importance of glacier ice as an archive of the E.S. processes has been more clearly recognized in a broad spectrum of scientific disciplinary studies [Langway, 1970; Dansgaard et al., 1973] involved in earth and planetary sciences. Precipitation is deposited on polar glaciers in a continuous sequence. Under the load pressure of subsequently deposited snowfall, the annual layers sink to greater depths, compress and get thinner. At a depth of 50 to 70 m in Greenland, the porous firn (consisting of ice grains with approximately spherical shape due to sintering) closes off, entrapping atmospheric air in the existing void spaces. Thus, the air bubbles in old glacier ice contain samples of atmospheric gases

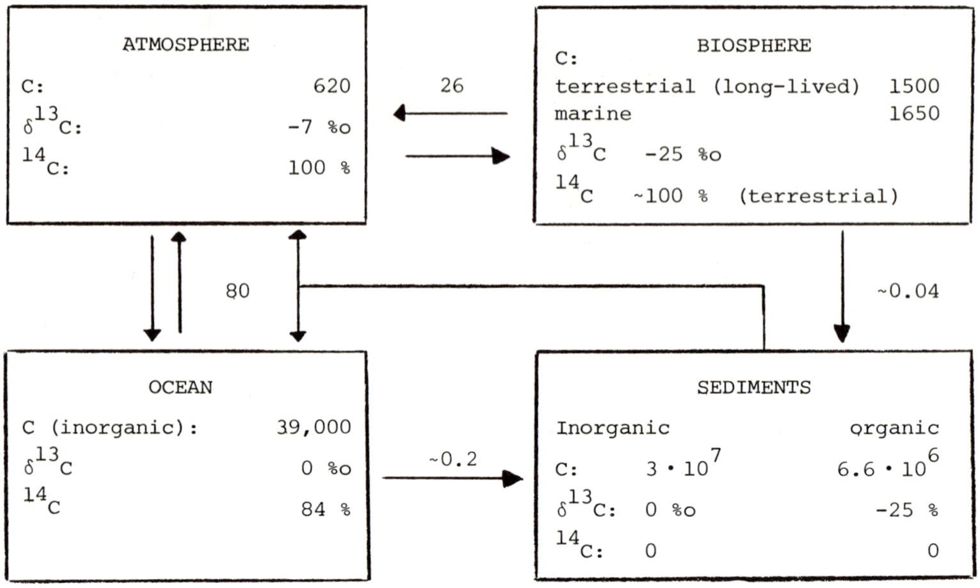

Fig. 3. The amounts of carbon are in 10^{15} g C, and fluxes in 10^{15} g C/a. The isotope values given are average concentrations. The exchange flux biosphere–atmosphere refer to the long-lived terrestrial biota only. The ^{14}C concentrations indicated are corrected for isotope fractionation to a common $\delta^{13}C = -25‰$. The actual uncorrected atmospheric and oceanic ^{14}C concentrations are 103.6% and about 88%, respectively, of modern wood ^{14}C.

existing at the time of pore close-off [Stauffer, 1981]. This unique entrapment of atmospheric gases is found in no other natural archive.

Of special interest in the context of human impact on the environment is the global carbon system (CO_2, CO, CH_4, C, . . .) and its uptake capacity for anthropogenic CO_2. Also of great importance is the terrestrial radiation and energy balance system and its sensitivity to anthropogenic interaction (CO_2 and trace gases, dust, albedo change, etc.). In what follows we show how the E.S. studies on the natural polar glacier archive contribute to the understanding of the CO_2 and the climate system.

The CO_2 System in Polar Glaciers and Other Natural Archives

The global carbon system consists of the atmosphere (mainly CO_2, CO, C, CH_4, . . .), the biosphere (organic C), the oceans (CO_2, HCO_3^-, $CO_3^=$, organic C), and the sediments (carbonates and organic materials) (Figure 3). The atmospheric CO_2 exchanges mainly with the carbon in the biosphere and the ocean. During periods of changing environmental conditions the exchange fluxes between the different reservoirs need not cancel out or balance, furthermore, the atmospheric CO_2 content (Na) may change:

$$\frac{dN_a(t)}{dt} = -F_{ab}(t) + F_{ba}(t) - F_{ao}(t) + F_{oa}(t) \qquad (1)$$

where, Fik(t) equals the exchange fluxes between reservoirs atmosphere (a), biosphere (b), and ocean (o).

During periods of major climatic change, the biosphere may grow or shrink in size and extract CO_2 from the atmosphere and indirectly from the ocean, or conversely release CO_2 into these systems. Furthermore, the atmospheric CO_2 concentration is essentially determined by the CO_2 partial pressure (pCO_2) in the ocean-surface water. The pCO_2 may change due to changes in ocean size and temperature as, for example, during glacial-interglacial transitions [Broecker, 1982]. However, pCO_2 changes may also be caused by changes in ocean circulation that would lead to chemical property changes in the surface waters relative to the deep ocean water [Oeschger et al., 1984]. Meaningful and significant changes in the atmospheric CO_2 content (N_a) are directly observable by analyzing the gas content of the air occluded in glacier ice [Scholander et al.,

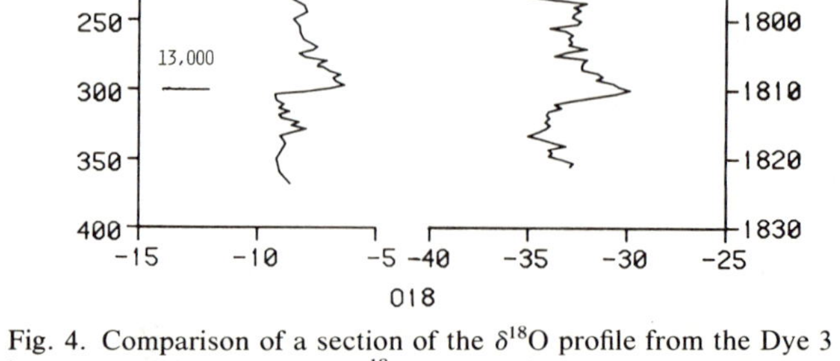

Fig. 4. Comparison of a section of the $\delta^{18}O$ profile from the Dye 3 ice core (right) with the $\delta^{18}O$ record in lake carbonate from Gerzensee (left). The strong similarities suggest that both records represent the same sequence of climatic events and thus the same time period.

1961; Berner et al., 1980; Raynaud and Delmas, 1977].

Measurements of the isotopic ratios ($^{13}C/^{12}C$ and $^{14}C/^{12}C$) allow the resolution of whether the atmospheric CO_2 changes are caused by either an exchange with the biosphere or hydrosphere. Figure 3 shows all the carbon reservoirs along with the quantity of carbon and isotopic information contained in these reservoirs. Due to a relatively rapid exchange, the radiocarbon in the biosphere is approximately equal to that in the atmosphere ($R_b \simeq R_a$), whereas that in the mixed surface ocean layer is only about 95% of the radiocarbon in the atmosphere. On the other hand, the $^{13}C/^{12}C$ ratio ($\delta^{13}C$) of the atmosphere is closer to that of the ocean than to most of the plants (biosphere). A CO_2 increase in the atmosphere due to a net CO_2 flux from the biosphere to the atmosphere would therefore parallel a decrease in the $^{13}C/^{12}C$ ratio but without a significant change in the $^{14}C/^{12}C$ ratio. Conversely, if the atmospheric CO_2 increase were due to a net influx from the ocean, the $^{14}C/^{12}C$ ratio would decrease and the $^{13}C/^{12}C$ ratio would remain essentially constant.

Historical and older $^{14}C/^{12}C$ and $^{13}C/^{12}C$ ratios can be reconstructed by isotope measurements made on tree rings, corals, peat, and CO_2 extracted from glacier ice. Using the new accelerator mass spectrometer (AMS) enables a $^{14}C/^{12}C$ determination to be made on samples from as little as 1 cm^3 of CO_2, whereas the earlier analytical method of counting radioactive decay required at least 100 cm^3 of CO_2. The AMS is especially important in the investigation of the $^{14}C/^{12}C$ ratios in glacier ice samples. The new method can be applied to samples of only 10 to 30 kg of glacier ice, whereas in using the older method of radioactive decay of ^{14}C nuclei, up to 1 ton or more of ice was required.

Besides the internal changes in the carbon system, as discussed above, the atmospheric $^{14}C/^{12}C$ ratio can vary due to changes in the production of ^{14}C by cosmic radiation [Suess, 1968; Stuiver and Quai, 1980]. One is, therefore, compelled to pose the question: is it experimentally possible to decide whether ^{14}C variations measured in samples of tree rings, peat bogs, and sediments and in the CO_2 extracted from ice are primarily caused by external forcing (variations in production rates) or by an internal process (carbon system dynamics)? This question can indeed be answered: in addition to ^{14}C, the AMS makes it possible to measure the cosmic ray produced ^{10}Be and ^{36}Cl isotopes contained in only 1 kg of ice. These isotopes become attached to aerosols and are deposited with them. The residence time in the atmosphere of ^{10}Be and ^{36}Cl is relatively short; a matter of months to a few years. Changes in their production rates are therefore relatively unattenuated and are reflected in the glacier deposit with good time resolution. In contrast, cosmogenic ^{14}C is oxidized to $^{14}CO_2$ and diluted in the exchanging carbon system.

Comparison of the ^{14}C and ^{10}Be variations helps to answer the questions related to the external and internal causes of ^{14}C variations. Measurements of the ^{10}Be and ^{36}Cl variations will also serve as a check for the ^{14}C production rate variations. As a first approximation one can assume that ^{10}Be and ^{36}Cl are proportional to the ^{14}C production variations ($P_{14_C}(t)$). Based on tree-ring measurements, the atmospheric ^{14}C variations ($^{14}C(t)/C$) of the past 8000 years are known and from measurements of CO_2 extracted from ice cores it is possible to get information on atmospheric $^{14}C/C$ variations over a longer time range. $P_{14_C}(t)$ and atmospheric $^{14}C(t)/C$ are related to each other via the response function of the CO_2 system to a ^{14}C pulse-input, $R(\tau)$; we obtain the following relation:

$$^{14}C(t)/C = {}_0\int^{\infty} P_{14_C}(t - \tau)\, R(\tau) d\tau \quad \text{[Oeschger, 1982]}$$

In short, measurements of ^{14}C and ^{10}Be variations will reveal a) the history of the atmospheric isotope production rate by cosmic radiation in the form of ^{10}Be variations; b) information on the response $R(\tau)$ of the CO_2 system to atmospheric ^{14}C production variations, and c) information on variations in the carbon cycle with time where ^{14}C variations are observed without indications of triggering ^{14}C production rate variations based on ^{10}Be data.

The Climate System Information from Ice Cores

The average global energy balance on the earth's surface is given by:

$$S\pi R^2(1 - A) = 4\pi R^2 \sigma T_S{}^4(1 - B)$$

With S = solar constant
R = radius of earth
A = albedo (reflected fraction of solar irradiation)
σ = Stefan-Boltzmann constant
B = fraction of infrared radiation (emitted from surface), absorbed in atmosphere and reemitted back to surface
T_S = surface temperature

Variations of the parameters S, A and B lead to changes in the passive parameter T_S, the earth's surface temperature. Information on all of these four parameters are recorded in ice cores.

Variations in the emission of solar plasma lead to changes in the magnetic shielding of galactic cosmic radiation in the inner part of the solar system, producing variations in the cosmic ray flux reaching the earth. They are reflected as variations in the production of radioactive nuclei in the earth's atmosphere. As already mentioned, these production rate changes are recorded in tree rings as $^{14}C/^{12}C$ ratio changes and in precipitation (rain or snowfall), as changes in the cosmogenic radioisotopes ^{10}Be and ^{36}Cl. It seems plausible that changes of other solar properties like luminosity and ultraviolet emission which might influence climate are somehow related to the solar plasma emission changes. Changes in the ^{10}Be content in ancient deposited snowfalls therefore might also reflect changes in solar parameters (like *S*) which could influence climate.

Variations in atmospheric turbidity which alters the earth surface's albedo *(A)* are induced by volcanic eruptions and lead to stratospheric dust layers. Solid electrical conductivity measurements on ice cores enable the identification of volcanic dust and therefore contribute to the reconstruction of the history of atmospheric turbidity [Hammer et al., 1980]. The information obtained from solid electrical conductivity measurements is enhanced by more detailed ionic chemistry measurements made on adjacent samples that identify concentration levels in chemical species and allows a more complete interpretation of the resulting data [Herron and Langway, 1985].

Air bubbles in the ice constitute samples of ancient atmospheres. Ice seems to be the only natural archive in which essentially undisturbed air samples are stored. Measurements of the gaseous composition reveal variations of the composition and concentrations of infrared active gases like CO_2 and CH_4 [Craig and Chou, 1982], which influence parameter *B*.

Finally, changes in temperature *(T_S)* are revealed as changes in the $^{18}O/^{16}O$ ratio of the water molecules. Water slightly depleted in ^{18}O relative to sea water reflects warm periods and water strongly depleted in ^{18}O reflects cold periods. Under favorable conditions, the high resolution of the ice core stable isotopic information enables reconstruction of seasonal ^{18}O variations as far back as at least 10,000 years. In addition, elevated air temperatures (above 0°C) lead to surface melting, and the thickness of melt layers can be used as indicators of the length and intensity of the summer melting periods [Herron et al., 1981]. At locations on ice caps where the mean annual temperature is $\leq -30°C$, melting of ice is negligible or nonexistent, although solar radiation crusts are found.

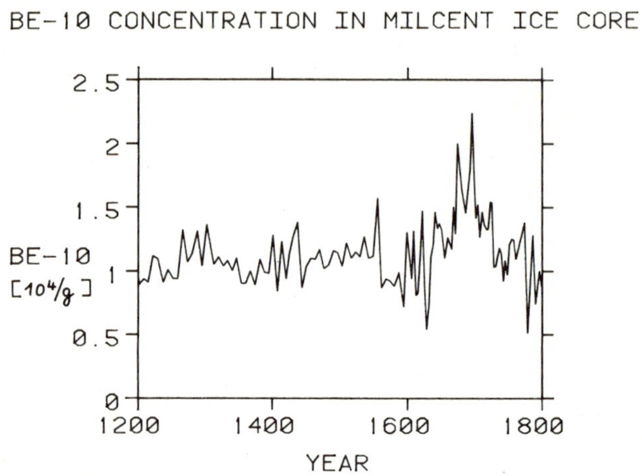

Fig. 5. ^{10}Be concentration in units of 10^4 atoms/g ice as a function of age in the Milcent ice core (Greenland). The age was determined based on annual variations of $\delta^{18}O$ [Hammer, 1978; Dansgaard and Johnsen, personal communication; Beer, 1983b].

Results

In the following we present results partly available at the AGU/GISP Meeting, June 1982, but partly obtained only during the last 18 months, where some of the proposed projects already turned out very valuable, partly unexpected data. We mainly concentrate on observations which illustrate the E.S. concept outlined in this paper.

1. The Temperature and Climate Information

First we briefly discuss the $\delta^{18}O$ profile measured for the Dye 3 ice core [Dansgaard et al., 1985], since it provides the most valuable information on the climatic evolution and helps to arrange in proper order all the other ice core data. The $\delta^{18}O$ variations are especially representative for the polar regions where the core is obtained but also reflect significant climatic events of hemispheric or even global significance. Most of the pronounced $\delta^{18}O$ oscillations from the Wisconsin stage ice age and the Wisconsin-Holocene transition observed in the Dye 3 core can be correlated with the oscillation observed in the Camp Century core. The pronounced $\delta^{18}O$ variations at the Glacial-Postglacial transition can also be observed in the carbonate deposits of Central European lake sediments [Oeschger et al., 1984; Dansgaard et al., 1984.]. Some of the rapid $\delta^{18}O$ oscillations reflecting relatively warm periods in the generally cold Wisconsin stage may also have their counterparts in warmer periods in Central Europe as reflected in pollen profiles studied in peat bogs or lake sediments as suggested in Figure 4. The $\delta^{18}O$ variations as found in the Greenland ice sheet cores and in Central European lake sediments will provide the framework and basis for all of the E.S. information on the climatic evolution of the North Atlantic Ocean areas.

2. The Atmospheric CO_2 History

The CO_2 measurements of the air occluded in very cold ice ($< -30°C$), which is not influenced by melting and refreezing, shows that between 1000 AD and 1800 AD the atmospheric CO_2 content was between 260 ppm to 280 ppm [WMO, 1983]. This is to compare with an atmospheric value of 315 ppm in 1958 (at the beginning of the period of accurate atmospheric CO_2 measurements) and 340 ppm measured in 1982. The ice core data therefore strongly support the hypothesis that the presently observed increase of atmospheric CO_2 is an anthropogenic phenomenon [Oeschger and Stauffer, 1983].

CO_2 measurements on Dye 3 ice-core samples from the Glacial-Postglacial transition [Stauffer et al., 1985] confirm the observation found earlier on Greenland and Antarctic ice cores [Berner et al., 1980; Delmas et al., 1980; Neftel et al., 1982] that at the end of the Wisconsin the CO_2 concentration was 180 to 200 ppmv and then increased at the transition, parallel to the $\delta^{18}O$ shift to values in the 260 to 300 ppmv band. Stauffer et al. [1985], and Oeschger et al. [1984] discuss the possibility that the main CO_2 increase already occurred during the first warming period, around 13,000 BP.

Detailed CO_2 concentration measurements in the period 40,000 BP to 30,000 BP in the Dye 3 core reveal that also during the Wisconsin the atmospheric CO_2 concentration showed rapid changes [Stauffer et al., 1984]. They will be discussed later together with the oscillation of other parameters.

3. ^{14}C Variations in Tree Rings and ^{10}Be Variations in Ice

(Information on the response of the CO_2 system to an atmospheric perturbation.)

Accelerator mass spectrometry has rendered measurements of the cosmic ray produced isotope ^{10}Be on ice core samples of 1 kg possible.

The ^{10}Be measurements on samples covering the past 80 years correlate well with sun spot numbers [Beer et al., 1983a]. Periods with low solar activity correspond to periods with relatively high ^{10}Be concentrations in the ice. These results are in agreement with direct observations made by neutron monitors during the last few solar cycles. It was shown by the monitors that low solar activity means relatively little shielding of the galactic component of cosmic radiation reaching the inner part of the solar system and thus a relatively high rate of radioisotopes in the earth's atmosphere is produced. During periods of high solar activity, the ^{10}Be concentrations were correspondingly lower.

The ^{10}Be measurements on ice core samples covering the past 800 years show relatively high concentrations (factor 1.6) during the

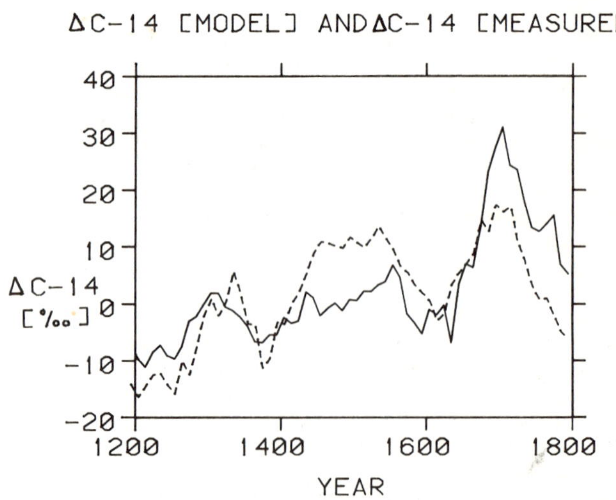

Fig. 6. $\Delta^{14}C$ in the atmosphere. Dashed curve: tree-ring measurements [Stuiver and Quay, 1980]; solid curve: $\Delta^{14}C$ model-calculated based on production rates determined from the ^{10}Be data [Beer, 1983b].

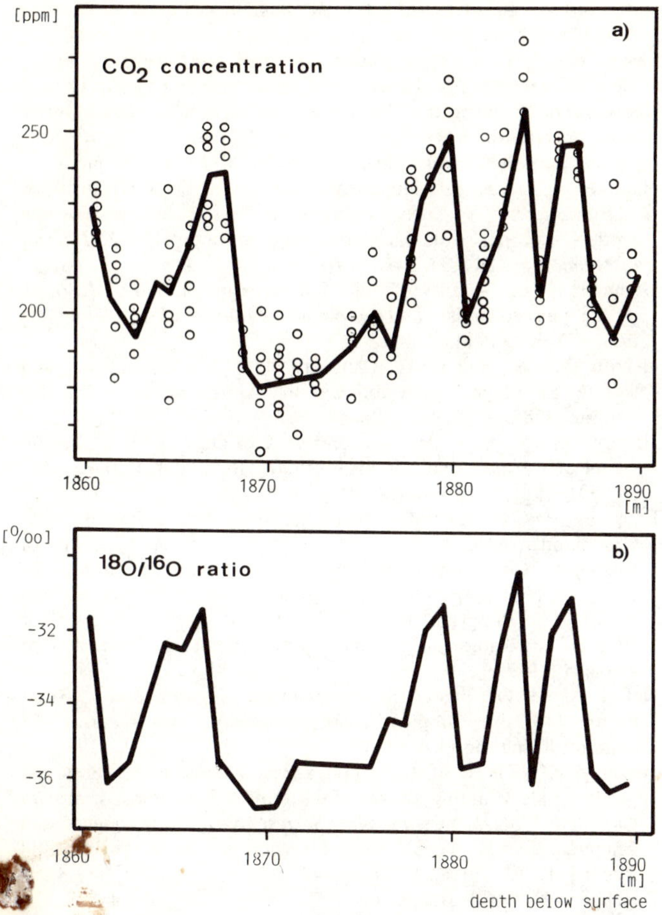

Fig. 7. CO_2 and $\delta^{18}O$ values measured on ice samples from Dye 3 (the 30-m increment corresponds to about 10,000 years, see Figure 4). a) Circles indicate the results of single measurements of the CO_2 concentration of air extracted from ice samples. The solid line connects the mean values for each depth. b) The solid line connects the $\delta^{18}O$ measurements done on 0.1-m core increments.

period of the quiet sun from 1640 to 1700 AD, the so-called Maunder Minimum of solar activity (Figure 5) [Beer et al., 1983b]. This relatively high ^{10}Be concentration parallels an increase of the $^{14}C/^{12}C$ ratio found in tree rings [Stuiver and Quay, 1980]. If we assume that the relative ^{14}C production variations are proportional to the relative ^{10}Be variations, the expected ^{14}C variations can be predicted by model calculations. For the Maunder Minimum period the calculated ^{14}C variations agree well with the measured values strongly suggesting that the recent ^{14}C variations are indeed the result of ^{14}C production rate variations and supporting the validity of the carbon cycle models used to calculate disturbances in the E.S. (Figure 6).

4. During the Wisconsin the Ice Core Parameters Oscillate Between Two Value Bands

(A bistable climatic system.)

Ice core parameters from 40,000 to 30,000 BP [Dansgaard et al., 1985] show strongly fluctuating values. As examples, the CO_2 concentrations (Figure 7) [Stauffer et al., 1984] are plotted versus $\delta^{18}O$ values. The correlation between CO_2 and $\delta^{18}O$ is almost perfect; warm periods (indicated by high $\delta^{18}O$ values) show CO_2 concentrations of 250 to 270 ppmv, and cold periods (indicated by low $\delta^{18}O$ values) values between 180 and 200 ppmv. ^{10}Be shows for the warm periods relatively high values, similar to those observed for the Holocene. During the cold periods, the ^{10}Be concentrations were higher by a factor of 1.5 to 2.5. Similar changes of other ice core parameters, like dust and chemical composition, are reported by Dansgaard et al. [1984], and Finkel and Langway [1984] for this time period.

Parameters like $\delta^{18}O$, CO_2, ^{10}Be, solid conductivity, and chemical composition thus give vivid descriptions of the state of the E.S. Several transitional changes between a cold and warm E.S. are observed. The cold state is characterized by low $\delta^{18}O$ and CO_2, and a high solid conductivity and chemical constituents. As is shown, these parameter indicators shift from cold to warm signals, in tandem series, within a century. Oeschger et al. [1984] proposed that the main element which provides the difference in the two states is ocean circulation. In the cold state, the North Atlantic Ocean is not producing deep water as it is in the Holocene, and the world ocean in general was in a relatively stagnant state. Little heat is transported to high latitudes, and large parts of the North Atlantic Ocean are covered by sea ice. Upwelling is at a minimum, and thus the supply of the ocean surface with CO_2-rich water is limited. Consequently, the average CO_2 partial pressure in the ocean surface water is low, corresponding to the relatively low atmospheric CO_2 content. Temperature and humidity of the north-polar region and also Greenland are low, as well as the snow accumulation rates which in turn corresponds to a higher concentration of the cosmic ray produced radioisotope ^{10}Be. The poleward heat transport is caused mainly by the strong winds which interact intensively with the agitated ocean surface and captures sea salts (and chemical constituents). At the same time, unconsolidated continental dust combines with the sea spray, and both components are transported to high latitudes. Based on the climate model, the low CO_2 concentrations correspond to global temperatures which were about 1.5°C lower than during the warmer state. Over ice-covered regions the difference was probably even higher by about a factor of 2 to 3. The warm state then would correspond to a circulating ocean bringing heat to the North Atlantic Ocean and the surrounding areas, and forming deep ocean water. The sea ice cover would be relatively small. The upwelling process would bring CO_2-rich water to the ocean surface layer increasing its average CO_2 partial pressure and the atmospheric CO_2 concentration. Temperature would be relatively warm with high humidity and therefore the ^{10}Be would be diluted. The latitudinal gradients and the winds were less and therefore the production and transport of marine and continental dust would be smaller.

It seems that during the milder part of the Wisconsin stage both the cold and warm states were more or less equally important. During the coldest part the cold state dominated and at the Wisconsin/Holocene transition the entire system made a final shift to the warm state, which we are still in.

This discussion could be carried much further. Questions such as: What were the boundary conditions which enabled this bistable climate system? What mechanism triggered the changes from one state to the other? In addition to low accumulation, do the high ^{10}Be concentrations also reflect periods of less solar magnetic shielding combined with a reduced solar output?

5. Separation of Natural and Anthropogenic Climatic Change Based on Ice Core Information

In conclusion, as has been discussed earlier in this paper, ice cores contain all of the essential data on factors forcing climatic change (solar properties, volcanic dust, infrared-active gases, etc.) and on parameters describing the resulting new climatic state ($\delta^{18}O$, marine and continental dust, melt features, etc.). The climatic data can be obtained on a bipolar basis by further investigating ice cores from both Greenland and Antarctica. This information would systematically advance our knowledge on deterministic climatic processes for the last 100 to 1000 years and longer. This knowledge would then be of great help to separate the CO_2-induced climatic warming from the background of climatic variations caused by solar parameter changes and atmospheric dust load. Such a concerted research effort conducted on ice cores could bring about a decisive breakthrough regarding one of the most urgent problems of man's future—the climate.

Acknowledgments. The author expresses his gratitude to all the motivated individuals who contributed to the success of the Greenland Ice Sheet Program: in the program planning, the development and construction of the drilling and analytical equipment, the drilling operation, the analytical work in the field and in the laboratory, and in data reduction and interpretation. This research was possible only through the many years of support by the U. S. National Science Foundation, the Swiss National Science Foundation, the Danish Commission for Scientific Research in Greenland, and the U. S. Department of Energy.

References

Beer, J., M. Andrée, H. Oeschger, B. Stauffer, R. Balzer, G. Bonani, C. Stoller, M. Suter, W. Wölfli, and R. C. Finkel, Temporal ^{10}Be variations in ice, *Radiocarbon*, *25* (2), 269–278, 1983a.

Beer, J., U. Siegenthaler, H. Oeschger, M. Andrée, G. Bonani, M. Suter, W. Wölfli, R. C. Finkel, and C. C. Langway, Jr., Temporal ^{10}Be Variations, Proceedings of the 18th International Cosmic Ray Conference, Bangalore, India, August, 1983, Vol. 9, pp. 317–320, 1983b.

Bennett, C. L., R. P. Beukens, H. E. Gove, R. B. Liebert, A. E. Litherland, K. H. Purser, and W. E. Sondheim, Radiocarbon dating using electrostatic accelerators. Negative ions provide the key. *Science*, *198*, 508–510, 1977.

Berner, W., H. Oeschger, and B. Stauffer, Information on the CO_2 cycle from ice core studies, *Radiocarbon*, *22*, 227–235, 1980.

Broecker, W. S., Ocean chemistry during glacial time, *Geochimica et Cosmochimica Acta*, *46*, 1689–1705, 1982.

Craig, H., and C. C. Chou, Methane: the record in polar ice cores, *Geophysical Research Letters*, *9* (11), 1221–1224, 1982.

Dansgaard, W., S. J. Johnsen, H.B. Clausen, and N. Gundestrup, Stable isotope glaciology, *Meddelelser om Grønland*, *197*, 53, 1973.

Dansgaard, W., H. B. Clausen, N. Gundestrup, C. U. Hammer, S. J. Johnsen, P. M. Kristinsdottir, and N. Reeh, A new Greenland deep ice core, *Science*, *218*, 1273–1277, 1982.

Dansgaard, W., S. J. Johnsen, H. B. Clausen, D. Dahl-Jensen, N. Gundestrup, C. U. Hammer, and H. Oeschger, North Atlantic climatic oscillations revealed by deep Greenland ice cores, American Geophysical Union, *Geophysical Monograph 29*, *Climate Processes and Climate Sensitivity* (Maurice Ewing Series), 288–298, 1984.

Dansgaard, W., H. B. Clausen, N. Gundestrup, S. J. Johnsen, and C. Rygner, Dating and climatic interpretation of two deep Greenland ice cores, this volume, 1985.

Delmas, R. J., J. M. Ascencio, and M. Legrand, Polar ice evidence that atmospheric CO_2 20,000 y BP was 50% of present, *Nature*, *284*, 155–157, 1980.

Finkel, R. C., and C. C. Langway, Jr., Temporal variations in the chemical composition of ice cores as related to CO_2, $\delta^{18}O$, and dust records during the Wisconsin (abstract), Chapman Conference on Natural Variations in Carbon Dioxide and the Carbon Cycle, Tarpon Springs, Florida, January 9–13, 1984, in press.

Hammer, C. U., H. B. Clausen, W. Dansgaard, N. Gundestrup, S. J. Johnsen, and N. Reeh, Dating of Greenland ice cores by flow models, isotopes, volcanic debris, and continental dust, *Journal of Glaciology*, *20* (82), 3–26, 1978.

Hammer, C. U., Acidity of polar ice cores in relation to absolute dating, past volcanism and radioechoes, *Journal of Glaciology*, *25* (93), 359–372, 1980.

Herron, M. M., and C. C. Langway, Jr., Chloride, nitrate, and sulfate in the Dye 3 and Camp Century, Greenland ice cores, this volume, 1985.

Herron, M. M., S. L. Herron, and C. C. Langway, Jr., Climatic signal of ice melt features in Southern Greenland, *Nature*, *293*, 389-391, 1981.

Langway, C. C., Jr., Stratigraphic analysis of a deep ice core from Greenland, *Research Report 77*, 130 pp., U.S. Army Corps of Engineers, Cold Regions Research and Engineering Laboratory, Hanover, N.H., 1967 (also *Geological Society of America Special Paper 125*, 186 pp., 1970).

Muller, R. A., Radioisotope dating with a cyclotron, *Science*, *196*, 489–494, 1977.

Neftel, A., H. Oeschger, J. Schwander, B. Stauffer, and R. Zumbrunn, Ice core sample measurements give atmospheric CO_2 content during the past 40,000 y. *Nature*, *295*, 220–223, 1982.

Oeschger, H., The contribution of radioactive and chemical dating to the understanding of the environmental system, *American Chemical Society Symposium Series*, *No. 176*, Nuclear and Chemical Dating Techniques: Interpreting the Environmental Record, L. A. Currie, Editor, 5–42, 1982.

Oeschger, H., and B. Stauffer, Review of the history of the atmospheric CO_2 recorded in ice core, Sixth Oak Ridge National Laboratory Life Sciences Symposium on The Global Carbon Cycle, Knoxville, Tennessee, October 31–November 2, 1983, in press.

Oeschger, H., J. Beer, U. Siegenthaler, B. Stauffer, W. Dansgaard, and C.C. Langway, Jr., Late-glacial climate history from ice cores, American Geophysical Union, *Geophysical Monograph 29*, *Climate Processes and Climate Sensitivity* (Maurice Ewing Series), 299–306, 1984.

Raynaud, D., and R. Delmas, Composition des gaz contenu dans la glace polaire, symposium on Isotopes and Impurities in Snow and Ice (Proceedings of the Grenoble Symposium, 1975), *IAHS-AISH Publication No. 118*, 377–381, 1977.

Scholander, P. F., E. A. Hemmingsen, L. K. Coachman, and D. C. Nutt, Composition of gas bubbles in Greenland icebergs, *Journal of Glaciology*, *3*, 813–822, 1961.

Stauffer, B., Mechanismen des Lufteinschlusses in natürlichem Eis, *Zeitschrift für Gletscherkunde und Glazialgeologie*, *17*, 17–56, 1981.

Stauffer, B., H. Hofer, H. Oeschger, J. Schwander, and U. Siegenthaler, Atmospheric CO_2 concentration during the last glaciation, Proceedings of the Symposium on Ice and Climate Modelling, Evanston, Illinois, June 27–July 1, 1983, *Annals of Glaciology*, *5*, 160–164, 1984a.

Stauffer, B., A. Neftel, H. Oeschger, and J. Schwander, CO_2 concentration in air extracted from Greenland ice samples, this volume, 1985.

Stuiver, M., and P. D. Quay, Changes in atmospheric carbon-14 attributed to a variable sun, *Science*, *207*, 11–19, 1980.
Suess, H. E., Climatic change, solar activity, and the cosmic-ray production rate of the natural radiocarbon. *Meteorological Monographs*, *8*, 146–150, 1968.
Suter, M., R. Balzer, G. Bonani, and W. Wölfli, Radioisotope dating using an EN-tandem accelerator. *IEEE Transaction on Nuclear Science*, *NS-28* (2), 1475–1477, 1981.
WMO, Report of the WMO (CAS) meeting of experts on the CO_2 concentration from pre-industrial times to I.G.Y., WMO Project on Research and Monitoring of Atmospheric CO_2, *Report No. 10*, WCP-53, 1983.

A BATTERY POWERED, INSTRUMENTED DEEP ICE CORE DRILL FOR LIQUID FILLED HOLES

N. S. Gundestrup and S. J. Johnsen[1]

Geophysical Isotope Laboratory, University of Copenhagen, Denmark

Abstract. The ice core drill used at Dye 3 for coring to bedrock at 2038-m depth is described. The main design criteria was low cost and weight, and easy maintenance in the field. The drill is divided into two main parts. The upper section consists of the antitorque section that prevents rotation of the upper part, the motors and the electronics. During drilling, the ice chips produced by the cutters are sucked into the lower, rotating part of the drill and transported with the drill to the surface, where the drill is clamped to a 6 m tower and tilted to horizontal position for easy core removal and drill cleaning. The cutters work like a plane, with an aggressive cutting angle that reduces the cutting power and provides stable penetration essentially independent of the load on the cutters. The drill is powered by a rechargeable battery pack, charged through the 6.4-mm cable. A microcomputer built into the drill controls 7 functions (battery charge and temperature, motor rotation speed and direction, etc.) at the same time that it measures 32 parameters. All this information is transmitted to a data terminal at the surface. The length of the drill is 11.5 m and its weight is 180 kg. The tower and the winch, including an electro-hydraulic pumpstation and 2500 m of cable, weigh 900 kg total. Core length is about 2.2 m per run, and the weekly core retrieval is 120 m of 10 cm diameter core at the 2000 m depth. The core recovery is better than 99.9%. Close to bedrock the hole deviates 6 degrees from the vertical, and the temperature is −13°C (−20°C at surface).

Introduction

In the 2037.63-m long core from Dye 3 in South Greenland (65°11′N, 43°49′W, h=2490 m), every snowfall in the last 40,000 years is preserved in an unbroken sequence together with impurities and a small amount of air. In order to retrieve such core, a new deep ice core drill, ISTUK (IS= ice in Danish, TUK= drill in Greenlandic), was developed. Previously, 2 deep (exceeding 1000 m) ice cores to bedrock existed. One from Camp Century (1966) in northwest Greenland [Ueda and Garfield, 1968], the other at Byrd Station (1968), Antarctica [Ueda and Garfield, 1969]. The drill used to recover these cores was lost in 1969 [Garfield and Ueda, 1976], and no deep ice core drill was then available. An oil-rig type drill, the wire-line system [Hansen, 1976], has been under development, but further work on this drill was terminated in 1978. In recent years another thermal drill for use in liquid filled holes has been made in the USSR [Zotikov, 1979], however the operating principles of this drill limit the applicability of the core.

Our aim with the ISTUK drill system was to provide continuous ice cores with as few core breaks as possible. The ice core is dated by counting annual layers in the core. Given an annual layer thickness of, for example, 8 cm, if just a few centimeters of core are missing it could influence the accuracy of the dating. A thermal drill was not used because the thermal shock creates stress in the core introducing additional breaks. In addition, a thermal drill consumes about 5 kW of power [Mellor and Sellmann, 1976], and transport of this amount of energy to a drill up to 3500-m downhole is difficult. The resultant decision to use a probe-type electromechanical system, with the drill suspended on the cable, was based on experience with the Danish shallow drill [Johnsen et al., 1980].

The ISTUK drill

The drill is shown in Figure 1. The length of the drill is about 11.5 m, and it bores a hole 129.5 mm in diameter. The core diameter is 102.3 mm. The ice is cut with a plane-like system consisting of 3 aggressive knives backed by stopper shoes that control the pitch or depth of cut. The produced cuttings, "chips", mix with the hole fluid and the created slush is sucked into 3 channels on the outside of the drill with each channel ending in a piston pump chamber.

The upper part of the drill, including the antitorque system, pressure chamber with motors, and a screw at the lower end of the pressure chamber, is prevented from rotating by 3 leaf springs pressed against the hole wall. The cable terminates in a weight located inside the antitorque system in such a way that allows free rotation of the cable at the drill. The weight can be used as a hammer in order to create an impact to break the core.

The shaded parts in Figure 1, i.e. the triangular shaft, the piston rod and 3 pistons, are attached to the motor exit shaft that extends through the hollow screw. The triangular shaft provides rotation of the drill barrel through the linear ball bearing. At the same time, the roller nut on top of the drill barrel creates a lengthwise movement of the rotating drill barrel thus changing the distance between the pistons fixed to the rod and a set of discs fixed to the drill barrel. This movement creates changes in the volume of the chip chambers, resulting in a piston suction pump action. The air that is released from the ice and trapped in the sucking system is vented through ball valves.

The drilling proceeds in what are typically called "runs", and every run consists of the following steps. (1) First the drill is lowered into the hole. (2) When the drill touches the bottom, the lower part of the drill rotates, cutting away an annulus of ice. The produced chips are sucked into chip chambers inside the drill. (3) After drilling about 2.2 m of core, the rotation is automatically terminated and a pull on the cable activates the core catchers that break the core. (4) The drill with ice core and chips is hoisted to the surface where the drill is clamped to the tower and brought to a horizontal position for

[1] Now at: Science Institute, University of Iceland, Reykjavik, Iceland.

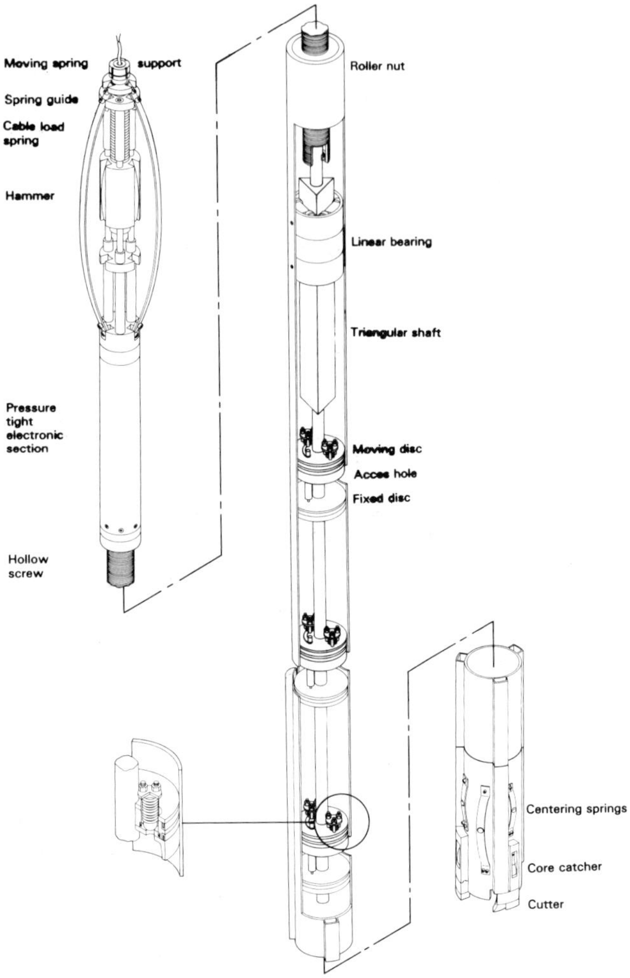

Fig. 1. ISTUK drill. The drill is suspended on a cable. The 3 leaf springs in the antitorque section prevent rotation of the upper part of the drill. The core is cut with the 3 knived drill head, lower right, and the cuttings are sucked into the drill. The leaf springs at the drill head keep the cutters free of the casing.

easy core removal and drill cleaning. The tower is hinged like that used in the Danish shallow drill [Johnsen et al., 1980]. After removal of the slush and rewinding of the drill, the cycle is repeated. At a depth of 2000 m, one run takes approximately 2 hours.

The upper part of the hole was cased by U.S. Army Corps of Engineers, Cold Regions Research and Engineering Laboratory (CRREL) [Rand, 1980b]. In order to counteract the hydrostatic pressure of the ice, the hole is filled with a mixture of JET A1 jet fuel and perchlorethylene (PCE).

Power Consumption

The size of the drill cable and thereby the dimensions of the winch depend on the power needed to operate the drill several kilometers down the borehole. The power required to drill had been estimated to be 500 W (based on Johnsen et al. [1980]) which is considerably lower than is consumed in the Byrd drill. The 500 W are consumed during drilling alone, and since the drilling time is of the order of 10% of the total time, the rest is spent in hole transit and at the surface where the average power consumption is approximately 50 W. A Ni-Cd battery pack in the drill provides storage of energy for the drilling, which makes it possible to use a small diameter cable. The battery pack consists of 55 2 Ah "SAFT" commercial C-size cells heated to 20°C. The overall power consumption of the drill, including heating of the batteries is only 100 W. This power consumption is so low, that only the breaking strength of the cable determines its size. Thus the 6.4 mm cable used can power a drill 3500 m down the borehole.

The winch is of the electrohydraulic type with a capability of regulating the speed continuously from −1 m/s to +1 m/s within ±0.02 m/s. The maximum pull is 8000 N. The total weight of winch and 2500 m of cable is less than 1000 kg.

Control and Communication

A drill has to be controlled from the surface. In conventional designs, the drill motors are switched ON and OFF by a switch at the surface, and each sensor in the drill has at least one separate wire in the cable. In the current design, the switch and sensor wires are replaced with digital AC information riding on top of the DC battery charging current in the single conductor coaxial type drill cable. The communication is in the form of CCITT audio tones between a modem (modulator-demodulator) in the drill and another at the surface. Thus the drill is controlled using a normal computer terminal.

The drill electronics are placed inside a pressure-tight chamber able to withstand the 180 bars pressure in the hole. One of the major advantages of a computer in the drill is that it simplifies the drill operation and improves the reliability. Thus the operator sends commands about only rotation speed and direction etc., while the drill computer takes care of routine control, including battery charging, discharging, temperature, etc. In addition, the drill computer works as a logger, and it measures cutterload, inclination and voltages. It detects and reacts to conditions that might endanger the drill, i.e. insufficient antitorque, leakage through the high pressure sealings, battery outgassing, etc.

The charging of the battery is treated with special care. Information from the manufacturer [Hodge et al., 1975] indicates that the lifetime of the batteries is several thousand charge-discharge cycles, if the following conditions are observed: The battery temperature should be close to 20°C; the batteries should never be deeply discharged, nor overcharged; and charging current should be between 1 amp and 4 amp. Most of these requirements are easy to fulfill. Overcharging is prevented by the drill computer when the

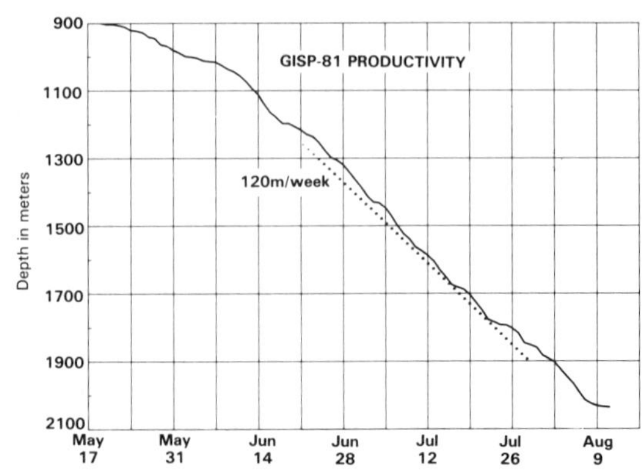

Fig. 2. 1981 production curve, depth vs. time.

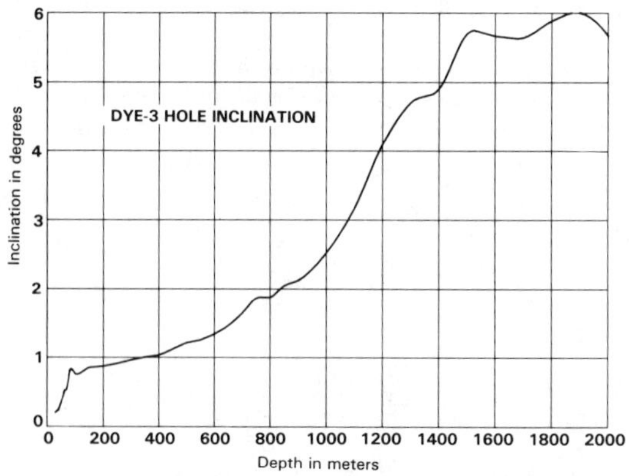

Fig. 3. Hole inclination. Azimuth is not known.

information on battery temperature, current, and voltage indicate full recharge. This method is so effective, that no battery failures occurred during the entire operation.

Results

The idea of making a new drill occurred in the autumn of 1977. The basic principles were tested at Dye 3 in the summer of 1978, and a prototype was tested at CRREL in the spring of 1979 [Rand, 1980a]. In 1979 the casing was installed by CRREL at Dye 3 [Rand, 1980b], and subsequently the prototype drilled to 225 m. The drilling proceeded slowly due to imperfections in the drill and the casing. In 1980 the drill was improved, but a mechanical breakdown terminated the season at a depth of 901 m. In 1981 drilling started with run 755 and, after the initial adjustments, it proceeded smoothly with a mean production rate of 120 m per week (Figure 2). The Wisconsin/Holocene transition was found at 1786.8-m depth, and the first trace of bottom material was encountered in run 1371 at a depth of 1949.45 m. Silty ice appeared in run 1400 at 2012.83 m, and the final depth was 2037.63 m in run 1418 when the drill became entrapped in the hole. The drill was recovered in good condition in 1982.

The ice core is of excellent quality, and no part of the core is known to be missing. It is estimated that less than 2 m of core was lost, resulting in a core recovery better than 99.9%. The brittle zone (the depth interval in which the core becomes brittle when hoisted to the surface) was 700 m to 1400 m, compared with 400 m to 900 m at Byrd station [Ueda and Garfield, 1969].

The specific cutting energy, the energy used to produce 1 m^3 of cuttings, increased with depth in spite of improvements in the cutting system. Close to bottom the specific energy was about 16 MJ/m^3. This increase agrees with an estimated increase in fracture stress with pressure [Shoji and Higashi, 1978]. From the Holocene/Wisconsin transition downwards, "the viscosity" of the ice decreased considerably [Shoji and Langway, 1985], but the specific energy was not affected.

A cable-suspended electromechanical drill tends to deviate from the vertical. Figure 3 shows the hole inclination with depth. At a depth of 1400 m, modifications of the cutters stabilized the inclination at 6 degrees compared with 15 degrees at bedrock in the Byrd hole [Ueda and Garfield, 1969]. The azimuth is not known.

The temperature close to the surface was −20°C and at the bottom the transducers inside the drill measured the temperature to −12°C in 1981. In 1982 a calibrated thermistor measured the hole temperature close to bottom at −13.40°C.

Conclusion

A new deep ice core drill for cold ice has been developed. It has a demonstrated capability to drill a 2 km deep ice core, and the same design may be used to a depth of 3.5 km. With minor modifications of the electronics, the drill will probably operate at temperatures as low as −50°C. The penetration rate is 120 m per week, and the drill is maintainable in the field. The ice core is continuous and of good quality. Due to the use of a microprocessor in the drill, an operator may be trained in just 3 days.

Acknowledgments. The drill head used in 1980 and 1981 was manufactured at the University of Bern by Mr. Henri Rufli. The mechanical workshops at the H.C. Oersted Institute, and the Niels Bohr Institute made several components in addition to providing field assistance. Jakob Schwander served as a drill team leader. Numerous colleagues from the universities of Bern and Copenhagen served as drill operators. The Polar Ice Coring Office, University of Nebraska-Lincoln, provided logistic support, and the 109th TAG, New York Air National Guard and the Royal Danish Air Force provided air support. Danish Arctic Contractors helped in Sondre Stromfjord, and the personnel at the Dye 3 radar station displayed unique hospitality and patience.

This work has been sponsored by the U.S. National Science Foundation, Division of Polar Programs, under contract DPP 8004667; the University of Bern; the University of Copenhagen; the Danish Commission for Scientific Research in Greenland; and the Danish Natural Science Research Council.

References

Garfield, D. E., and H. T. Ueda, Resurvey of the "Byrd" station, Antarctica, drill hole, *Journal of Glaciology, 17*(75), 29–34, 1976.

Hansen, B. L., Deep Core Drilling in the east Antarctic Ice Sheet: A Prospectus, in *Ice Core Drilling,* edited by J. F. Splettstoesser, pp. 29–36, University of Nebraska Press, Lincoln, Nebraska, 1976.

Hodge, B. J. R., R. Bonnaterre and F. Putois, Fast Charging of Sealed Nickel-Cadmium Batteries: Theory and Practice. Proceedings of a Symposium, available from SAFT UK LTD, Castle Works Station Road, Hampton, Middlesex, England, pp. 211–231, 1975.

Johnsen, S. J., W. Dansgaard, N. Gundestrup, S. B. Hansen, J. O. Nielsen, and N. Reeh, A fast light-weight core drill, *Journal of Glaciology, 25*(91), 169–174, 1980.

Mellor, M., and P. V. Sellmann, General considerations for drill system design, in *Ice Core Drilling,* edited by J. F. Splettstoesser, pp. 77–111, University of Nebraska Press, Lincoln, Nebraska, 1976.

Rand, J., The Danish Deep Drill, Progress Report: February–March 1979, *Special Report 80-3,* U.S. Army Corps of Engineers, Cold Regions Research and Engineering Laboratory, Hanover, N. H., 1980a.

Rand, J., 1979 Greenland Ice Sheet Program, Phase 1: Casing operation, *Special Report 80–24,* U. S. Army Corps of Engineers, Cold Regions Research and Engineering Laboratory, Hanover, NH, 1980b.

Shoji, H., and A. Higashi, A deformation mechanism map of ice, *Journal of Glaciology, 21*(85), 419–427, 1978.

Shoji, H., and C. C. Langway, Jr., Mechanical properties of fresh ice core from Dye 3, Greenland, this volume, 1985.

Ueda, H. T. and D. E. Garfield, Drilling through the Greenland Ice Sheet, *Special Report 126,* U.S. Army Corps of Engineers, Cold Regions Research and Engineering Laboratory, Hanover, N. H., 1968.

Ueda, H. T., and D. E. Garfield, Core drilling through the Antarctic Ice Sheet, *Technical Report 231,* U.S. Army Corps of Engineers, Cold Regions Research and Engineering Laboratory, Hanover, N. H., 1969.

Zotikov, I. A., Antifreeze-thermodrilling for core through the central part of the Ross Ice Shelf (J-9 Camp), Antarctica, *Report 79-24,* U.S. Army Corps of Engineers, Cold Regions Research and Engineering Laboratory, Hanover, N. H., 1979.

ULTRASONIC VELOCITIES AND CRYSTALLINE ANISOTROPY IN THE ICE CORE FROM DYE 3, GREENLAND

Susan L. Herron[1], Chester C. Langway, Jr., and Keith A. Brugger

State University of New York at Buffalo, Department of Geological Sciences, Ice Core Laboratory, Amherst, New York

Abstract. Ultrasonic velocity measurements on polar ice core samples can be used to monitor the development of crystalline anisotropy through an ice sheet. In a single crystal of ice, P-waves travel about 170 ms^{-1} faster along the c-axis than along a-axes. A randomly-oriented, poly-crystalline aggregate of ice will appear accoustically isotropic. However, with depth in an ice sheet, as c-axes become preferentially oriented in the vertical direction, the vertical velocity in the polycrystalline aggregate increases, as does the velocity difference, V.

Ultrasonic velocities have been measured on 248 samples from 38 depth intervals on the 2037 m ice core from Dye 3, Greenland. These are the first measurements ever made on fresh, unrelaxed samples from a deep ice core. In the Dye 3 core, the velocity difference increases from near zero in the upper several hundred meters to about 80 ms^{-1} by a depth of 1700 m. There is an abrupt increase in V to values of over 120 ms^{-1} which occurs over a very short depth interval marking the Holocene-Wisconsin climatic transition. The ultrasonic velocity results obtained from the Dye 3 ice core will be compared to the measured fabric profile and to other physical properties measured throughout the core.

Introduction

The analysis of crystalline fabric in polar ice cores is generally pursued in order to assess the development of anisotropy throughout an ice sheet. Studies on the deep ice cores from Byrd Station, Antarctica [Gow and Williamson, 1976] and Camp Century, Greenland [Herron, 1982; Herron and Langway, 1982] have described the development of highly preferred crystal orientations with depth. In recent studies several investigators have experimentally demonstrated the intimate relationship between preferred orientations and ice deformation rates [Lile, 1978; Russell-Head and Budd, 1979; Baker, 1981; Duval, 1981]. The confirmation of this relationship emphasizes the importance of obtaining and quantifying orientation data from the new 2037 meter ice core from Dye 3, Greenland.

The most accurate technique for determining the fabrics of polycrystalline ice is to measure the orientations of individual crystals in a thin section using a Rigsby-type universal stage [Rigsby, 1951]. For the Dye 3 core this work has been completed at 23 depth intervals between 100 m and 1815 m. For each fabric analysis, several statistical parameters have been calculated in order to provide a quantitative description which can be used as a fabric factor in the development of flow models and can also be used to compare the degree of fabric development at various locations [Herron, 1982; Herron and Langway, 1982].

An alternate, more rapid, method of monitoring fabric development is to determine the ultrasonic anisotropy through the ice sheet by measuring ultrasonic velocities in vertical and horizontal directions on selected ice core samples [Bennett, 1972; Kohnen and Bentley, 1977; Kohnen and Langway, 1977; Gow and Kohnen, 1979; Kohnen and Gow, 1979]. In a single crystal of ice, P-waves travel approximately 170 m s^{-1} faster parallel to the c-axis than perpendicular to it [Bennett, 1972]. In contrast, in a polycrystalline aggregate of ice, if the individual crystals are randomly oriented the ice will appear to be acoustically isotropic. Within the ice sheet, if the c-axes become preferentially oriented in the vertical direction, the vertical velocity will increase, as will the velocity difference, ΔV, between the vertical and horizontal directions. The methods for determining ultrasonic velocities on ice core samples have been modified from those used in previous studies. Using the modified technique, ultrasonic velocities were measured at 38 depth intervals between 100 m and 1995 m in the Dye 3 core. The results are compared with the observed fabric patterns and with the fabric statistics.

Methods

Thin Section Analysis

During the 1980 and 1981 Dye 3 field seasons, vertical thin sections were prepared in the field laboratory within one week of core recovery, using the procedures outlined by Langway [1958]. In 1980 the initial sampling depth was 238 m, and sampling continued at approximately 100 m intervals down to 900 m. In 1981 the sampling started at a depth of 913 m and continued at 100 m intervals, with additional samples taken as time permitted, to a depth of 1995 m, only 40 m above the bottom. In all cases the thin sections were taken continuously over increments representing a minimum of one to three or more years' accumulation. Additional vertical thin sections were later prepared from the 1979 Dye 3 core at depths of 100 m and 160 m, and a single horizontal thin section was taken at 1815 m from the 1981 core.

Mean crystal intercept lengths were measured in both the horizontal and vertical directions on all thin sections. The results from all thin sections in a given depth interval were averaged and are reported here as average crystal diameters. At 23 selected depths c-axis orientations were measured on one to three thin sections using the procedures outlined by Langway [1958] and the refraction corrections supplied by Kamb [1962]. The orientation data from the vertical sections were rotated into a horizontal position and plotted on the lower hemisphere of a Schmidt equal-area net; in accordance with recent preference [Hooke et al., 1980], all the data are presented here in scatter diagrams rather than in contoured diagrams.

[1] Now at: Schlumberger-Doll Research, Ridgefield, Connecticut.

TABLE 1. Samples for Ultrasonic Velocity Measurements

Depth (m)	Total Length (cm)	# Samples	Vertical (cm)	Horiz. (cm)	Horiz. (cm)
101	34.2	6	5.7	5.7	—
161	36.0	6	6.0	6.0	—
238	93.6	18	5.2	4.5	3.7
301	76.5	17	4.5	4.8	4.0
327	35.4	6	5.9	6.0	—
350	35.4	6	5.9	5.9	—
405	48.0	6	8.0	5.2	4.3
450	35.4	6	5.9	5.9	—
481	36.4	4	9.1	5.0	—
550	35.4	6	5.9	5.9	—
551	29.0	5	5.8	5.9	—
605	81.4	22	3.7	4.6	3.2
695	11.4	3	3.8	6.5	5.2
751	29.5	5	5.9	5.9	—
805	30.8	7	4.4	4.8	—
852	33.6	6	5.6	5.6	—
900	25.0	5	5.0	6.5	—
912	31.0	5	6.2	9.5	5.9
994	24.4	4	6.1	9.3	4.6
1065	26.8	4	6.7	7.6	5.1
1141	22.4	4	5.6	5.7	5.4
1184	28.0	5	5.6	9.3	5.4
1280	29.5	5	5.9	8.8	5.7
1360	56.8	8	7.1	8.8	5.4
1411	34.5	5	6.9	8.6	6.0
1522	30.0	5	6.0	8.8	5.8
1589	93.8	14	6.7	8.9	5.6
1627	61.0	10	6.1	8.9	6.0
1699	59.0	10	5.9	9.3	5.9
1779	22.8	4	5.7	9.3	6.2
1788	19.6	4	4.9	9.3	6.1
1815	23.6	4	5.9	9.2	6.0
1836	19.6	4	4.9	9.3	6.0
1856	18.8	4	4.7	7.9	6.3
1882	23.2	4	5.8	9.3	6.1
1923	23.2	4	5.8	7.4	5.9
1950	23.6	4	5.9	9.3	6.3
1995	18.0	3	6.0	7.5	6.2

In order to quantify the orientation data, some statistical parameters have been calculated for each fabric analysis. The first is the length of the resultant vector, R, which is obtained from the summation of all measured c-axis orientations when each c-axis is treated as a unit vector [Mardia, 1972]. If all c-axes are oriented in the same direction, R will be equal to N, the number of c-axes. In this work R is normalized with respect to N, so the maximum possible value is unity. This statistic provides a measure of concentration about the mean direction and will be large for oriented samples.

The second statistical treatment applied to the Dye 3 fabrics is the calculation of eigenvectors and eigenvalues [Mardia, 1972] from each set of axial data. The eigenvectors are three orthogonal vectors which can be considered as the principal axes of the ellipsoid that best describe the axial distribution. The associated eigenvalues are equal to the lengths of the three semi-axes. The largest eigenvalue can be used to calculate a concentration statistic, $(E_3/N)^{1/2}$, which provides a measure of axial concentration about the mean direction [Hudleston, 1977]. The maximum value is unity, and high values are representative of preferred orientations.

The length of the resultant vector and the eigenvalue concentration statistic have been used previously to describe the development of preferred crystal orientations in the Camp Century ice core [Herron, 1982; Herron and Langway, 1982]. Further discussion on the calculation and use of statistical parameters for spherical data can be found in Mardia [1972].

The third form of data treatment used on the Dye 3 fabrics is the determination of the angle, α, which is the half-apex angle of the cone containing 90% of the c-axes in a given data set. Kohnen and Gow [1979] used this angle to describe the degree of clustering and to provide a quantitative means of comparison between fabrics and ultrasonic velocities. Values of α have also been reported for the Camp Century fabrics [Herron, 1982].

Ultrasonic Velocity Measurements

Vertical and horizontal ultrasonic velocities were measured on 248 samples from 38 depth intervals. At each depth, measurements were made on several continuous samples in order to increase the number of crystals measured at each depth and thus minimize the high degree of variability encountered in other studies where measurements were generally limited to one 7–10 cm sample from a given depth interval [Kohnen and Bentley, 1977; Kohnen and Gow, 1979]. In the upper 900 m of the core, ultrasonic samples were taken from three of the same depth intervals at which thin sections were prepared. Ultrasonic velocities were measured on these samples in the field within one to two weeks of core recovery and were remeasured in the laboratory about six months later. In each case the velocity difference, ΔV, was within 5 m s^{-1} of the value obtained from the field measurements. Additional ultrasonic measurements were performed between six and nine months following core recovery on samples taken from within a meter of thin section samples and also from intermediate depths. Below 900 m all ultrasonic measurements were performed at identical depths as thin section analyses, and all measurements were made within one to two weeks of core recovery.

The samples were cut by bandsaw to have three sets of parallel faces in order to make one vertical and two horizontal measurements. The total vertical distance and the individual sample dimensions are presented in Table 1. All sample surfaces were sanded smooth on a wire-mesh screen. Sample lengths were measured by calipers to 0.05 mm.

Ultrasonic velocities were measured using a Krautkramer-Branson USL 38 instrument with a 2.25 MHz transducer. This is a portable pulse-echo device capable of performing wall thickness measurements from one side of the test material. The procedure used for measuring P-wave velocity was to place the transducer on a test block and to calibrate the oscilloscope and digital readout to display the thickness of the block. For this study the standard was a 2 cm × 4 cm × 6 cm steel block which was factory calibrated for temperatures between 0°C and −35°C. A low viscosity silicone oil was used for the couplant. The transducer was placed on the prepared surface and the relative steel thickness of the ice was displayed. Ultrasonic velocities were calculated using the relationship

$$\frac{V_s}{V_i} = \frac{T_s}{T_i}$$

where V_s is the velocity in the steel block which is determined from the factory calibration, V_i is the velocity in the ice sample, T_s is the relative steel thickness displayed on the digital readout, and T_i is the measured sample thickness. The real sample dimensions and relative steel thicknesses were measured four times each.

The reproducibility of the method was tested on a sample taken from the Byrd Station ice core by repeating three sets of measure-

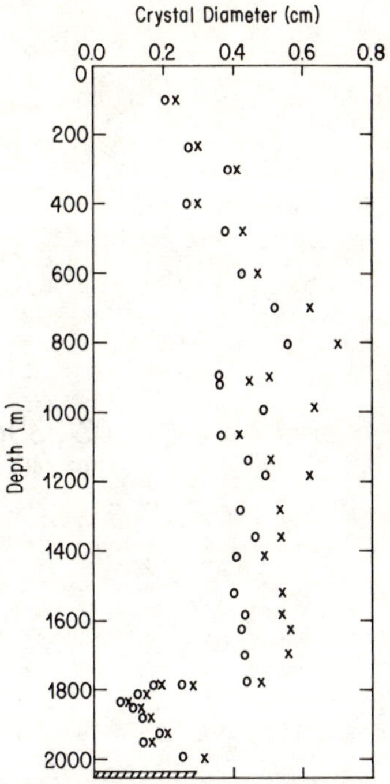

Fig. 1. Crystal size profile. The circles represent vertical dimensions, or intercept lengths, and the x's represent horizontal dimensions.

ments, one vertical and two horizontal, fourteen times in a period of about one month. These measurements yielded velocities of 3724 ± 17 m s^{-1}, 3819 ± 13 m s^{-1}, and 3775 ± 19 m s^{-1}, all of which have relative standard deviations of less than 0.5%. All Dye 3 ultrasonic velocities were corrected to +10°C using a correction of −2.3 m s^{-1} per °C [Kohnen, 1977].

Results

Crystal Size

The results of crystal size measurements are presented in Figure 1. In general, average crystal size increases gradually and the crystals become increasingly elongated in the horizontal direction between 100 m and 800 m. Below 900 m the crystals maintain a fairly constant size until about 1785 m where size decreases sharply. The depth interval at which crystal sizes decrease is the same interval that contains the Holocene-Wisconsin climatic transition [Dansgaard et al., 1982]. Below that, crystal sizes increase slightly.

For this study, crystal sizes are presented primarily to determine the average number of crystals represented in the ultrasonic velocity measurements. In the upper 900 m of the core the largest horizontal dimension in the ultrasonic samples is generally 6 cm (Table 1) which means that at 100 m a single horizontal measurement represents the average velocity through approximately 25 crystals, but at 800 m it represents the velocity through only about nine crystals. Below 900 m the largest horizontal dimension is generally nine centimeters, so from 900 m to 1700 m each ultrasonic measurement contains about 18 crystals. Below 1785 m the ultrasonic samples usually contain at least 45 crystals in a horizontal traverse.

Fabric Profile

The results of the Dye 3 fabric analysis are presented in Figures 2a (100 m–913m) and 2b (995 m–1815 m). In the upper 500 m of the core, the c-axes appear to be well distributed over the Schmidt net with no indications of preferred orientation. By a depth of 600 m there is a slight reduction in the number of near-horizontal c-axes, and below 800 m the c-axes are more definitely grouped in a broad, vertical cluster. Several fabrics between the depths of 805 m and 1183 m display poorly developed two-maxima patterns, generally with one maximum nearly centered on the vertical. A well-developed two-maxima pattern exists at 1360 m, but below this depth the strength of the nonvertical, weaker maximum rapidly diminishes. By a depth of 1588 m there is a very definite, strong vertical maximum, but it is accompanied by several weak clusters of nonvertical c-axes. Finally, in the deepest fabric at 1815 m, approximately 40 m below the climatic transition, there is a strong, single maximum that contains nearly all the measured c-axes and is oriented in a near-vertical position.

The only unusual feature in the fabric profile down to 800 m is the small circle pattern at a depth of 695 m which is not apparent in any of the adjacent sections. The only other depth at which a small circle pattern was obtained was 325 m which is not illustrated in Figure 2. The sample at 325 m was taken from a section of core that contains over 80% melt features, compared to a core average of only 5.6% [Herron et al., 1981]. The sample from 695 m contains almost 20% melt features while the melt-feature percentage in the other fabric samples is generally less than the core average. These and other preliminary analyses indicate that crystal orientations within melt layers frequently differ from surrounding fabrics, but there are not enough data to determine whether or not this could explain the fabric pattern observed at 695 m.

The statistics calculated for the Dye 3 fabrics indicate a trend of increasingly preferred orientations with depth. The normalized length of the resultant vector and the eigenvalue concentration statistic are two statistics that provide a measure of concentration about the mean direction; the profiles of these statistics are illustrated in Figure 3. Both values increase in a step-like fashion between 100 m and 1141 m with steps at 480 m and between 913 m and 995 m. Through this interval the lengths of the resultant vectors increase from around 0.69 to 0.80 and the concentration statistics from 0.74 to 0.83. Below 1141 m, the statistics increase in a less regular fashion until they attain their highest values ($R/N = 0.96$ and $(E_3/N)^{1/2} = 0.97$) for the deepest fabric analyzed at a depth of 1815 m. The statistics at 1279 m represent a fabric that is similar to the overlying one, but contains only 92 measured c-axes.

The half-apex angle, α of the cone containing 90% of the c-axes provides another measure of axial concentration. The α values for the Dye 3 core are presented in Figure 4. The decreasing values of α reflect the general trend of increasing fabric concentration with depth, but details of the profile differ slightly from those observed in Figure 3. The values display relatively small decreases below 405 m and again below 695 m with values changing from 78° to 69° to 61°. Below 1141 m the values are more erratic, but they decrease to about 45° and then to 23° for the fabric from 1815 m. Although all three sets of statistics reflect similar patterns of increasing orientation with depth, the changes occurring below 1411 m and at 1815 m are proportionately larger for α than for the other two statistics.

Ultrasonic Velocity Profile

The results of the vertical and horizontal ultrasonic velocity measurements are presented in Figure 5. In general, the horizontal velocities remain fairly constant with depth, showing only a slight increase from 3816 m s^{-1} in samples from 1183 m and above to 3826 m s^{-1} below 1183 m, and another increase to 3844 m s^{-1} below 1787 m. In contrast, the vertical velocity remains constant at about 3821 m s^{-1} in the upper 480 m and then increases almost linearly to 3864 m s^{-1} by 899 m. The next significant increase in vertical velocity is a jump to 3898 m s^{-1} occurring over the same depth

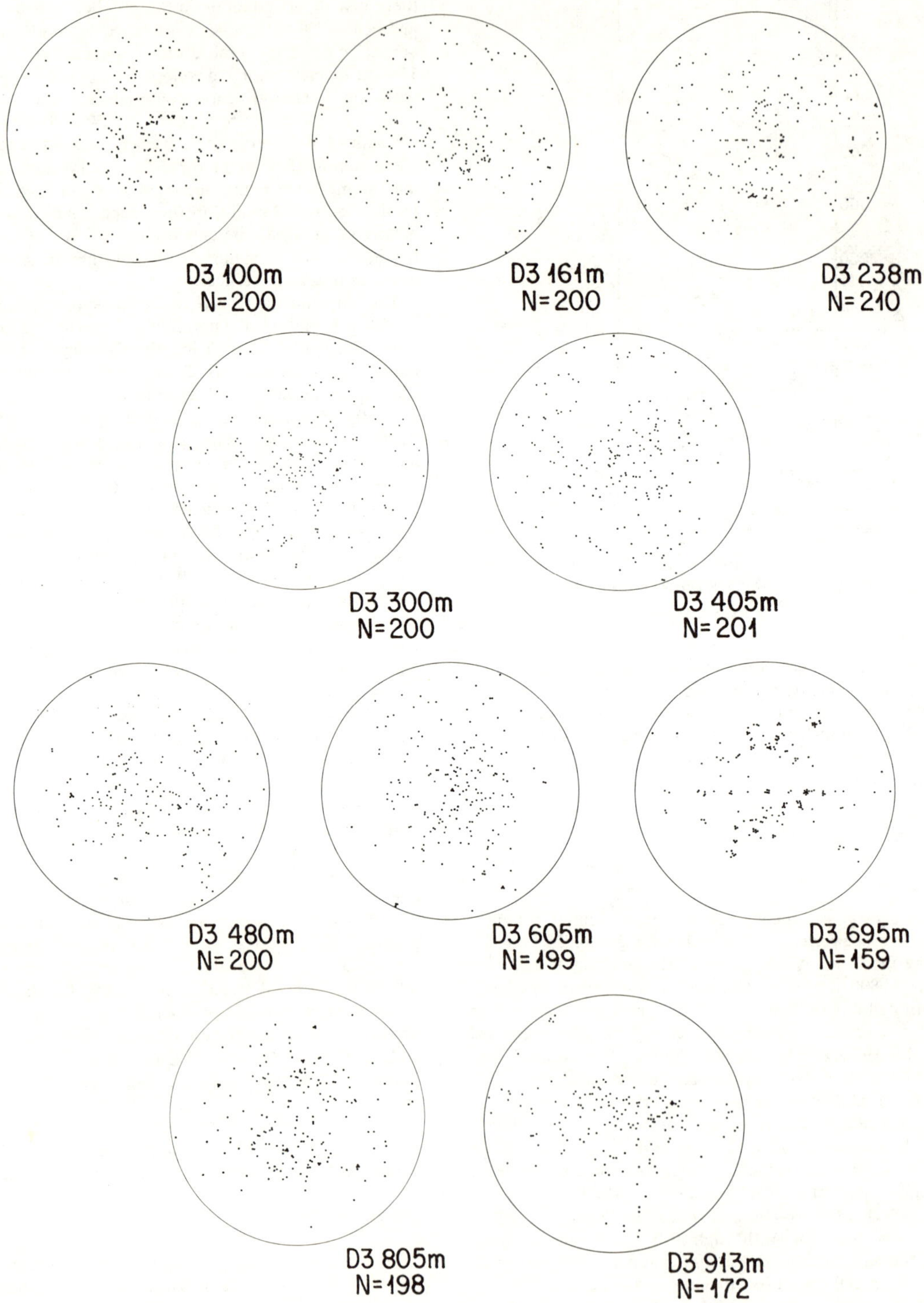

Fig. 2. a) Fabric profile (100 m–913 m). Dots represent c-axis orientations plotted on the lower hemisphere of a Schmidt net.

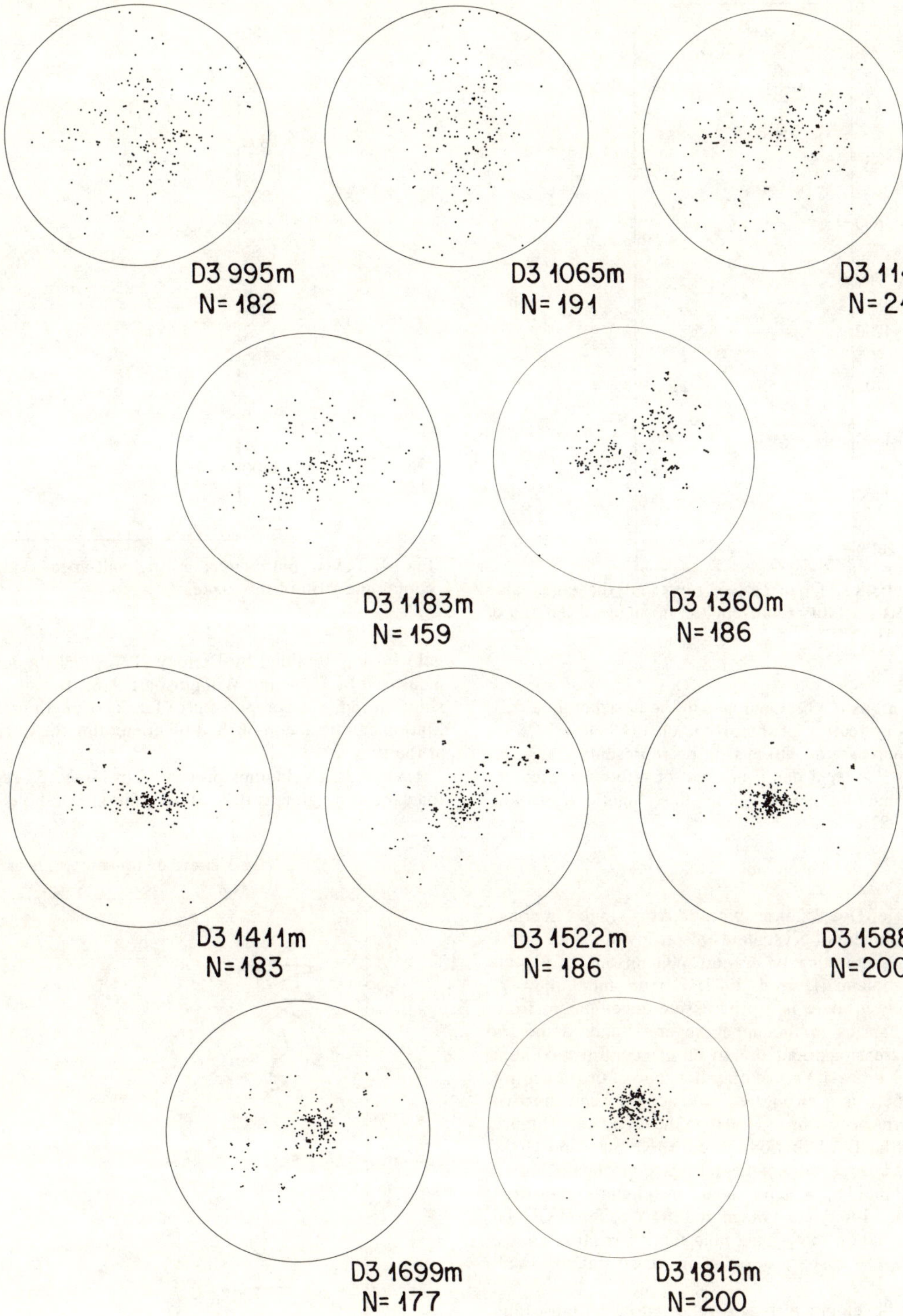

b) Fabric profile (995 m–1815 m).

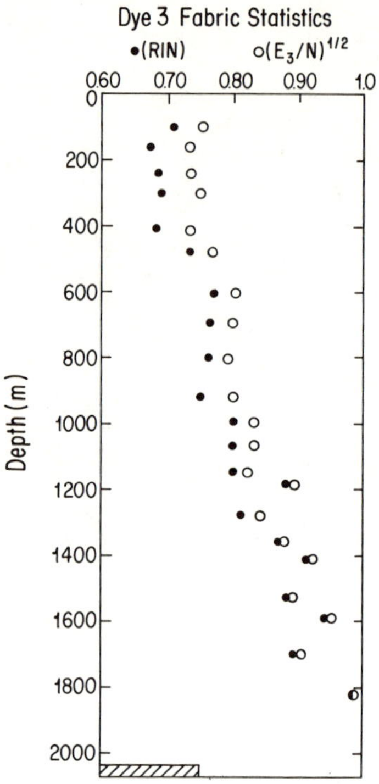

Fig. 3. Fabric statistics. Open circles represent the eigenvalue concentration statistics; dots represent the normalized lengths of the resultant vectors.

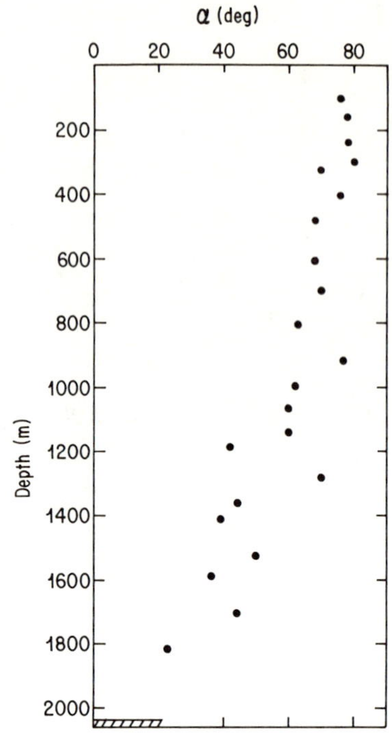

Fig. 4. Fabric parameter, α, the half-apex angles of the cones containing 90% of the c-axes.

interval (1183–1280 m) as the first increase in the horizontal velocities. The most striking feature of the profile occurs below 1778 m where the velocity increases to 3967 m s^{-1}; this represents a velocity increase of 167 m s^{-1} over a depth interval of only 10 m, and it occurs over the same depth interval as the climatic transition [Dansgaard et al., 1982].

Discussion

The strengthening of Dye 3 fabrics with depth is similar in some aspects to the development of crystalline anisotropy in the ice cores from Byrd Station, Antarctica [Gow and Williamson, 1976] and Camp Century, Greenland [Herron, 1982; Herron and Langway, 1982]. In all three cores there is a progressive development from randomly oriented fabrics in the upper several hundred meters through a zone of increasing orientation to a broad central maximum in moderately deep ice (~1000 m). Within the zone of broad central maximum fabrics there is a strong resemblance between the two maxima fabric patterns occurring at 1360 m in the Dye 3 core (Figure 2b) and 1137 m in the Byrd Station ice core [Kohnen and Gow, 1979]. Finally, in all three cores, the fabrics consolidate into a strong, single maximum at a depth near the climatic transition between the Holocene Interglacial Stage and the Wisconsin Glacial Stage. This transformation into single pole fabrics is fairly sharp, and it is accompanied by a sharp decrease in ice crystal size in all three cores.

The Dye 3 fabric development is also reflected in the ultrasonic velocity profiles presented in Figure 5, particularly the vertical velocity profile. The value of the vertical P-wave velocity profile as a qualitative indicator of fabric strength was demonstrated for the Byrd Station core using the continuous 75–1550 m ultrasonic velocity log obtained by Bentley [1972] and the ice crystal fabrics measured by Gow and Williamson [1976]. Because of this relationship, an indirect comparison of fabric strength at Dye 3 and Byrd Station can be accomplished by comparing the ultrasonic velocities at the two sites.

The vertical velocities presented in Figure 5 are generally lower than those obtained in the Byrd Station borehole [Bentley, 1972],

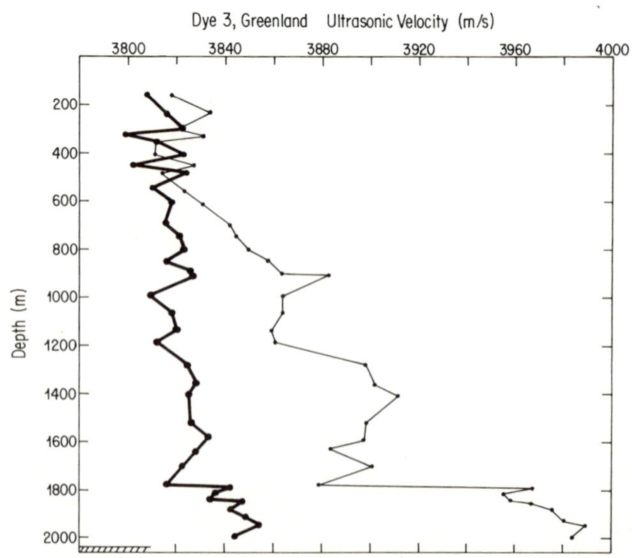

Fig. 5. Ultrasonic velocity profile. The thick line illustrates the horizontal velocities; the thin line represents vertical velocities.

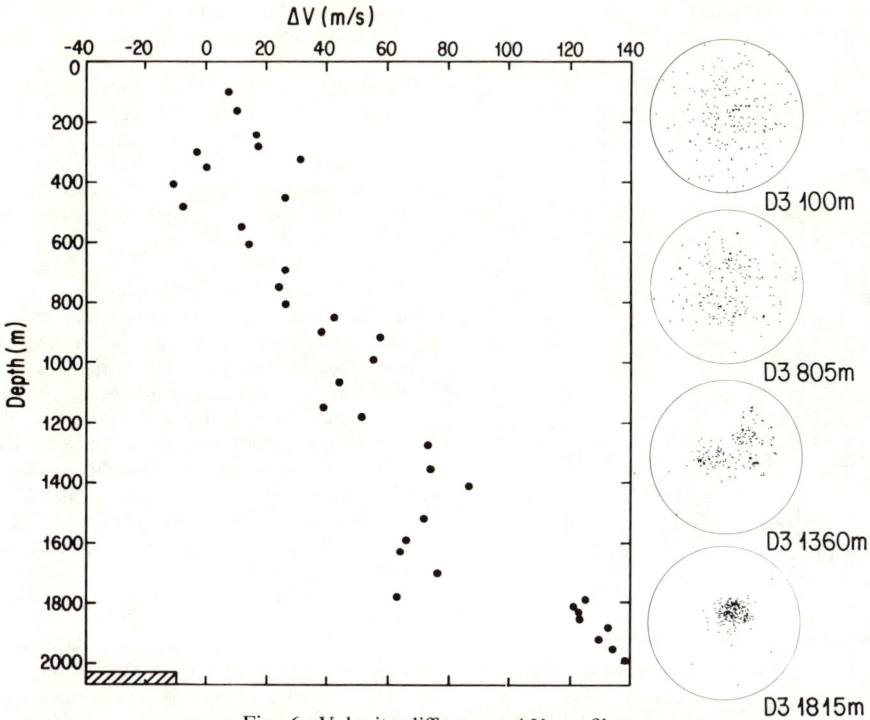

Fig. 6. Velocity difference, ΔV, profile.

but most of the difference can be accounted for by the temperature differences between the two sets of measurements. In the upper 100 to 1000 m of the two cores, where velocities tend to increase gradually with depth, the Dye 3 velocities are generally 20 to 55 m s^{-1} slower with an average difference of about 40 m s^{-1}. The Byrd Station borehole temperatures over this depth interval remain fairly constant at values around −27 to −28°C [Gow and Williamson, 1976], or 17 to 18°C colder than the temperature for which the Dye 3 velocities are reported. Using a temperature dependence of −2.3 m s^{-1} deg^{-1} [Kohnen, 1974], the velocity correction would be 39 to 41 m s^{-1} which is just equal to the observed differences between the two locations.

The jump in vertical velocities between 1180 and 1279 m in the Dye 3 core does not appear to have a direct counterpart in the Byrd Station velocity profile. This increase in the Dye 3 core does not appear to be related to any sudden change in crystalline orientations (see Figure 2), but it does occur over the same depth interval at which the air bubbles begin to disappear. The loss of discrete air bubbles below 1200 m is documented by records of visual stratigraphy and bubble pressure measurements (Herron, unpublished data), and below 1300 m it is documented by microscopic observations of what are believed to be clathrate hydrates [Shoji and Langway, 1982]. This absence of air bubbles in the deep ice can probably account for the increase in both horizontal and vertical ultrasonic velocities between 1180 and 1279 m in the Dye 3 core, but it does not explain the relatively higher increase of vertical velocities. Unfortunately, the effect of bubble disappearance on ultrasonic velocities cannot be evaluated on the other two deep ice cores. In the Byrd Station ice core the bubbles disappear very near the climatic boundary, at the same depth that the ice undergoes a significant textural change. In the Camp Century ice core the stratigraphic log indicates that air bubbles disappear between 1100 and 1165 m, which is also near the climatic transition, but the relaxed nature of the core made ultrasonic analysis impossible using the methods presented here [Herron, 1982].

The next stage at which the Dye 3 and Byrd Station ultrasonic velocities can be compared is in ice from the Wisconsin glaciation. This ice is represented in Dye 3 ultrasonic samples below 1787 m and in the Byrd Station ultrasonic log below 1200 m. The Dye 3 velocities average about 3985 m s^{-1} compared to 4015 m s^{-1} at Byrd Station. Using the temperature-velocity gradient of −2.3 m s^{-1} deg^{-1}, this velocity difference of 30 m s^{-1} corresponds to a temperature difference of −13°C. In fact, the temperatures for the Byrd Station velocities between 1200 and 1550 m are in the −25 to −20°C range, or −15 to −10°C colder than the Dye 3 samples, and consequently the measured ultrasonic velocities are essentially equivalent.

Overall, there is a remarkable similarity between the borehold ultrasonic velocities obtained by Bentley [1972] at Byrd Station and the vertical ultrasonic velocities obtained at selected intervals in the Dye 3 core. The close agreement between the two data sets suggests a similar development of crystalline anisotropy in the two cores.

This close agreement also demonstrates the importance of completing ice-core velocity measurements prior to ice-core relaxation. In an ultrasonic study conducted on the Byrd Station core about ten years following core recovery, Kohnen and Gow [1979] found that although they obtained excellent agreement with Bentley's borehole data above 1200 m, the vertical velocities they obtained for depths below 1200 m were substantially lower than the velocities measured in the borehole. They attributed the difference to cleavage crack formation and observed that the effect was greatest in the zone of strong axial fabrics. Similarly, vertical ultrasonic velocities in the Camp Century ice core were impeded by the existance of numerous cleavage cracks oriented in near-horizontal positions throughout the deep core [Herron, 1982].

Although the vertical velocity profile provides an indication of

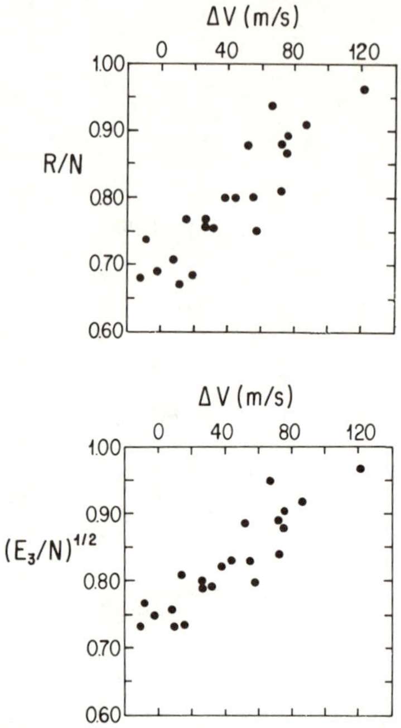

Fig. 7. a) Comparison of ΔV and R/N. b) Comparison of ΔV and $(E_3/N)^{1/2}$.

fabric development, the critical parameter is actually the velocity difference, ΔV, between the vertical and horizontal velocities [Bennett, 1972; Kohnen and Bentley, 1977; Kohnen and Langway, 1977; Kohnen and Gow, 1979]. The ΔV profile for the Dye 3 core is illustrated in Figure 6 along with selected fabric diagrams from four depth intervals.

The pattern that emerges from the ΔV profile is similar to those identified from the fabric profile and the fabric statistics. Above a depth of 500 m, values of ΔV range from -11 m s^{-1} to 31 m s^{-1} with no indication of a trend in fabric development. Below 500 m, the Dye 3 ΔV values increase steadily from 12 m s^{-1} to approximately 50 m s^{-1} by a depth of 1000 m. By 1279 m, ΔV increases to about 70 m s^{-1}, and this average is maintained for the next 500 m. Finally, at a depth of 1787 m, ΔV increases to 125 m s^{-1}, and it remains at values above 120 m s^{-1} through the deepest measurement at a depth of 1995 m where it reaches a value of 139 m s^{-1}.

The Dye 3 ΔV profile can be directly compared to the ΔV profile obtained from the Byrd Station core [Kohnen and Gow, 1979]. Over the top 1200 m, the two profiles are quite similar. At Byrd Station the upper 500 m of core yield velocity differences ranging from -13 to 60 m s^{-1} with no trend to indicate the development of crystalline anisotropy. Between 500 m and 1200 m the values tend to increase with depth with values between 25 and 99 m s^{-1}.

However, below 1200 m the two profiles differ considerably. In the Byrd Station core, ice between 1200 m and 1800 m is of Wisconsin age and it is expected that the ΔV values should be similar to those obtained below 1878 m in the Dye 3 core. Through this zone, the Byrd Station velocity differences vary erratically between values of 71 and 150 m s^{-1} with an average of only 102 m s^{-1}, whereas the Dye 3 velocity differences are higher and fairly constant with values between 121 and 139 m s^{-1}. Considering the fine-grained crystal sizes and highly oriented fabrics that characterize the Wisconsin ice in both cores, it appears that the Byrd Station velocity differences below 1200 m are affected by the problems of core relaxation and cleavage crack development.

In general, the whole ΔV profile from Byrd Station exhibits greater variability than that from Dye 3. The most likely cause for this is the difference in the sampling techniques and the number of crystals represented in each reported value of ΔV. In the study conducted by Kohnen and Gow [1979], ultrasonic travel times were measured on samples that were 10 cm long (vertical) and 7 cm in diameter (horizontal). The sample sizes were then reduced four or five times and travel times were remeasured with each size reduction in order to obtain a time-distance curve from which velocities were calculated. With each size reduction, the number of crystals was reduced, and consequently the results became less representative of the polycrystalline aggregate. In contrast, using the methods described earlier, the Dye 3 ultrasonic velocities were measured on several consecutive samples at each depth interval. At all depths except one the total vertical dimension was 20 cm or more, and in almost all cases horizontal velocities were measured on four or more adjacent samples (Table 1). The advantage of this sampling scheme is that it is possible to obtain a velocity difference that is much more representative of the ice mass at a given depth.

In order to directly compare the results of the fabric analyses with the ultrasonic velocity measurements, ΔV is plotted with the fabric statistics, R/N and $(E_3/N)^{1/2}$, in Figure 7. In both cases the values of ΔV increase linearly as the statistical parameters increase. The correlation coefficients between R/N and ΔV and $(E_3/N)^{1/2}$ and ΔV are 0.89 and 0.90, respectively. Both of these are significant at the 0.1% level.

The values of the cone angles containing 90% of the c-axes are plotted with ΔV in Figure 8. The correlation coefficient in this case is -0.80, which is also significant at the 0.1% level. The solid curve in this figure represents the values of ΔV that Bennett [1972] predicted for fabrics in which c-axes were evenly distributed within a solid cone of a given angle. In most cases the Dye 3 data plot above Bennett's curve with ΔV's frequently 20 to 40 m s^{-1} higher than the values predicted for a given cone angle. Similar results were obtained on the Byrd Station core by Kohnen and Gow [1979] who suggested that the cause for this lies in the fact that Bennett's curve is based on a uniform distribution of c-axes in the cone, while the natural distribution more closely approximates a Fisherian distribution. This seems to be the case for the Dye 3 fabrics as well; the fabrics within the 90% cone are still quite unevenly distributed, and it appears that a smaller cone angle would provide a better agreement with Bennett's predicted curve.

Conclusions

The development of crystalline anisotropy in the Dye 3 ice core is clearly evident in the fabric profile. In general, there is no indication of preferred orientations above 500 m, but below 500 m there is a steady reorientation of c-axes toward the vertical. Ice from the Wisconsin glaciation displays uniform fine-grained textures and strong single pole fabrics which are nearly identical to those observed in the ice cores from Camp Century and Byrd Station. The

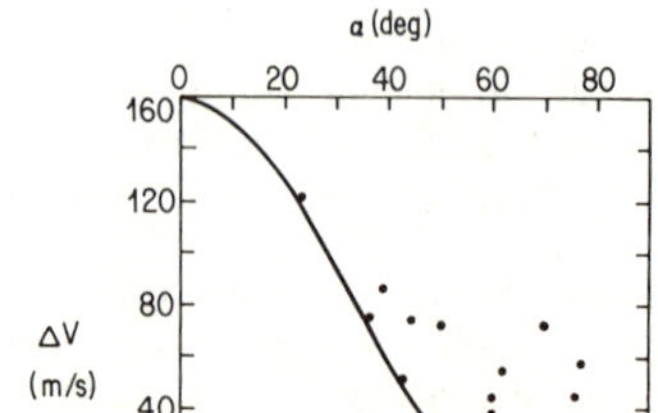

Fig. 8. Comparison of ΔV and α.

fabrics can be quantitatively described using either the length of the resultant vector or the eigenvalue concentration statistic. Both of these provide an exact measure of concentration about the mean direction, and they should prove to be useful fabric parameters in future flow models.

The development of crystalline anisotropy can also be described by the velocity difference, ΔV. The Dye 3 ΔV profile conforms quite well to the observed patterns as is demonstrated by the high correlation between fabric statistics and velocity differences. The sampling techniques used on the Dye 3 core and the measurement of fresh core samples both contribute to the success of the ultrasonic analysis.

The vertical velocity profile obtained from the ultrasonic measurements is remarkably similar to the borehole results obtained at Byrd Station. Furthermore, it is predicted that there will be excellent agreement between the results reported here and the results of future borehole logging at Dye 3. For this aspect of the work, the use of fresh core samples has been critical, especially in deep ice.

Acknowledgments. We thank R. F. Tompkins and M. L. Beeman for field and laboratory participation. This work was sponsored by the National Science Foundation, Division of Polar Programs.

References

Baker, R.E., Textural and crystal-fabric anisotropies and the flow of ice masses, *Science*, *211* , 1043–1044, 1981.

Bennett, H.F., Measurements of ultrasonic wave velocities in ice cores from Greenland and Antarctica, *Research Report 237*, U. S. Army Corps of Engineers, Cold Regions Research and Engineering Laboratory, Hanover, N. H., 1972.

Bentley, C.R., Seismic wave velocities in anisotropic ice: a comparison of measured and calculated values in and around the deep drill hole at Byrd Station, Antarctica, *Journal of Geophysical Research*, *77*, 4406–4420, 1972.

Dansgaard, W., H.B. Clausen, N. Gundestrup, and N. Reeh, A new Greenland deep ice core record reconciled with the Camp Century and deep sea records, *Science*, *218*, 1273–1277, 1982.

Duval, P., Creep and fabrics of poly crystalline ice under shear and compression, *Journal of Glaciology*, *27*, 129–140, 1981.

Gow, A.J., and H. Kohnen, The relationship of ultrasonic velocities to c-axis fabrics and relaxation characteristics of ice cores from Byrd Station, Antarctica, *Journal of Glaciology*, *24*, 147–154, 1979.

Gow, A.J., and T. Williamson, Rheological implications of the internal structure and crystal fabrics of the West Antarctic ice sheet as revealed by deep core drilling at Byrd Station, *Geological Society of America Bulletin*, *87*, 1665–1677, 1976.

Herron, M.M., S.L. Herron, and C.C. Langway, Jr., Climatic signal of ice melt features in southern Greenland, *Nature*, *293*, 389–391, 1981.

Herron, S.L., Physical properties of the deep ice core from Camp Century, Greenland, Ph.D. thesis, 153 pp. State University of New York at Buffalo, 1982.

Herron, S.L., and C.C. Langway, Jr., A comparison of ice textures and fabrics at Camp Century, Greenland and Byrd Station, Antarctica, *Annals of Glaciology*, *3*, 118–124, 1982.

Hooke, R. LeB., M. Mellor, W.F. Budd, J.W. Glen, A. Higashi, T.H. Jacka, S.J. Jones, R.C. Lile, R.T. Martin, Mechanical properties of polycrystalline ice: an assessment of current knowledge and priorities for research, *Cold Regions Science and Technology*, *3*, 263–275, 1980.

Hudleston, P.J., Progressive deformation and development of fabric across zones of shear in glacial ice, in *Energetics of Geological Processes*, edited by S. and S. Battacharji, pp. 121–150, Springer-Verlag, New York, 1977.

Kamb, B., Refraction corrections for universal stage measurements. I. Uniaxial crystals, *American Mineralogy*, *47*, 227–245, 1962.

Kohnen, H., The temperature dependence of seismic waves in ice, *Journal Glaciology*, *67*, 144–147, 1974.

Kohnen, H., and C.R. Bentley, Ultrasonic measurements on ice cores from Ross Ice Shelf, Antarctica drill hole, *Antarctic Journal of the U.S.*, 148–150, 1977.

Kohnen, H., and C. C. Langway, Jr., Untersuchungen mit ultraschall an eisbohrkernen aus Zentralgronland, *Polarforschung*, *47*, 1–10, 1977.

Kohnen, H., and A.J. Gow, Ultrasonic velocity investigations of crystal anisotropy in deep ice cores from Antarctica, *Journal of Geophysical Research*, *84*, 4865–4874, 1979.

Langway, C.C., Jr., Ice fabrics and the universal stage, *Technical Report*, *62*, U. S. Army Corps of Engineers, Snow, Ice and Permafrost Research Establishment, Wilmette, Ill., 1958.

Lile, R.C., The effect of anisotropy on the creep of polycrystalline ice, *Journal of Glaciology*, *21*, 475–484, 1978.

Mardia, K.V., *Statistics of Directional Data*, Academic, London, 1972.

Rigsby, G.P., Crystal fabric studies on Emmons Glacier, Mount Rainer, Washington, *Journal of Geology*, *49*, 590–598, 1951.

Russell-Head, D.S., and W.F. Budd, Ice-sheet flow properties derived from bore-hole shear measurements combined with ice-core studies, *Journal of Glaciology*, *24*, 117–130, 1979.

Shoji, H., and C.C. Langway, Jr., "Air hydrate inclusions in fresh ice core," *Nature*, *298*, 548–550, 1982.

MEASUREMENTS OF A KIND OF DC-CONDUCTIVITY ON CORES FROM DYE 3

A. Neftel, M. Andrée, J. Schwander[1], and B. Stauffer

Physics Institute, University of Bern, Switzerland

C. U. Hammer

Geophysical Isotope Laboratory, University of Copenhagen, Denmark

Abstract. The dc-conductivity measured with two electrodes that move slowly over an ice sample shows generally a good correlation with the acidity of the melted ice sample. The method has been suggested and successfully applied by C. Hammer. In this paper the experimental setup and some physical aspects of the method are discussed; results obtained on the Dye 3 core and their climatological interpretation are discussed in another paper in this volume by C. Hammer. With the newly developed measuring device the current is measured as a function of depth and recorded by an XY-recorder as well as in digitized form on magnetic tape. The resolution of the two recording methods are compared.

Based on observations made during the measurements at Dye 3 the effect of ice-core temperature on electrical conductivity is discussed. This serves as a contribution to the understanding of the basic physical phenomena involved in the electrical conductivity of ice. Finally some possible explanations for the observed seasonal variations are suggested and discussed.

Introduction

One of the goals of the deep drill operation at Dye 3 was to measure different physical and chemical parameters directly in the field. Most important are continuous records, which help to identify pronounced features in the core such as high acidity levels due to volcanic eruptions, the transition from Holocene to Wisconsin ice, and seasonal patterns. As an example, the formation of melt layers every summer at Dye 3 offers the possibility to count the years by visual stratigraphy, as long as the bubbles are visible [Herron, 1981]. Until now, rather sophisticated analytical procedures had to be performed to determine the levels and the variations of different impurities in ice cores from such remote stations as Dye 3.

The Electrical Conductivity Measurements (ECM), performed directly in the field, offer the possibility to monitor continuously the acidity of the core. This allows the identification of acidity originating from volcanic eruptions and the counting of seasonal variations in the Dye 3 core. If volcanic signals can be correlated with known historical eruptions, the corresponding layer can be absolutely dated. The ECM was first proposed by Claus Hammer [1980], and successfully applied on different ice cores such as Crête, Camp Century, and Byrd.

[1] Now at: State University of New York at Buffalo, Department of Geological Sciences, Amherst, New York.

In this paper we treat the methodological and physical aspects of the ECM and discuss possible causes for the seasonal patterns found in Greenland ice. Comparison with other continuous records such as dust, electrical conductivity, and pH measured in the liquid phase, as well as their climatological interpretation, are discussed elsewhere in this volume by Hammer et al.

Principles of the ECM

Two electrodes with an applied high voltage are moved over a flat ice surface and the current that flows between the electrodes is recorded. If the electrodes would rest on the ice the current would rise very fast and last a few tenths of a second decreasing rapidly afterwards. While moving the electrodes with a velocity between 5 cm sec^{-1} and 15 cm sec^{-1}, the current is not decreasing since the ice between the electrodes is replaced before the current starts to decrease. Therefore we measure this initial current as a function of the depth along an ice core. The main difference to a conductivity measurement in the liquid phase is that in ice protonic defects (H_3O^+ defects) contribute at least an order of magnitude more than the OH^- defects [Hobbs, 1974]. The measured current can be transformed into acidity by means of a calibration curve relating the two parameters in terms of μA measured (at a fixed temperature), and μ equivalents H_3O^+ derived from pH measurements on the melted samples. This calibration curve will be discussed.

Measuring System

The ECM has to be performed on a clean and fresh surface because the electrical properties in the surface layer change with storage time. This aging effect is discussed in detail in another paper [Schwander et al., 1983]. To provide a flat surface for the measurement, the cores were mounted in a frame and horizontally split with a bandsaw especially constructed for this purpose. After removing the smaller upper part, the lower part was cleaned by a microtome knife fixed on the saw. The fixation of the cores in a frame was extremely useful for brittle cores, originating from depths between 800 m and 1400 m. Without this fixation a continuous record would not have been obtained. Afterwards the cores were transported in the frame to the measuring platform in a room where the surrounding temperature could be kept below $-10°C$. The preparation of a flat, polished surface on the cores was also very useful for the visual stratigraphy. This enabled a detailed analysis of the annual percentage of melt layers down to a depth of 1200 m to be performed. Figure

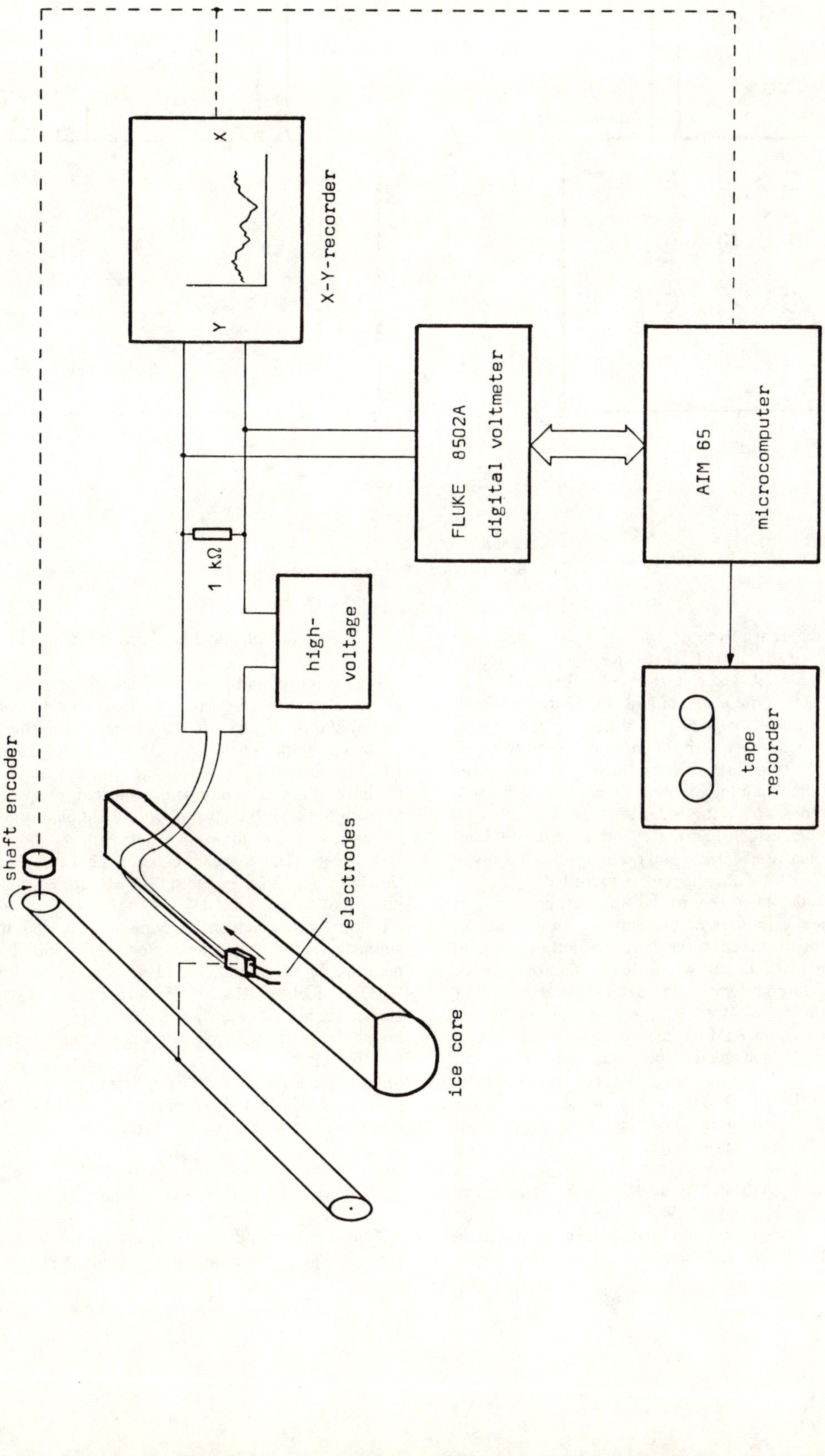

Fig. 1. Schematic diagram of the measuring system.

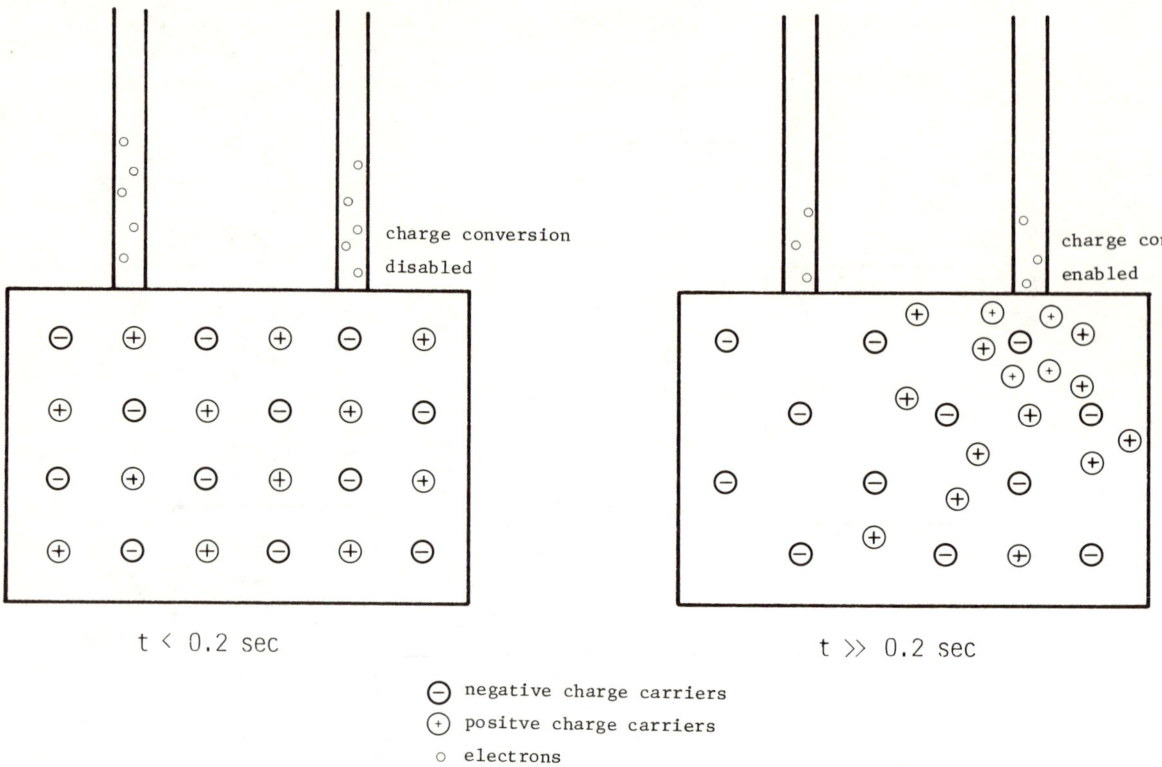

Fig. 2. Time dependence of charge transport in ice.

1 shows a schematic diagram of the measuring system used. The current was transformed by a 1 KΩ resistor into a potential difference which was recorded. The electrodes were connected to a motor-controlled carriage on the platform and guided by hand over the ice surface. To guarantee maximal reproducibility of the measurements and best comparability with the measurements of different cores from Dye 3, the same brass electrodes with a contact area of about 1.5 mm^2 and the same high voltage of 1250 V were used. The load on the electrodes was in the order of $5 \cdot 10^6$ Pa.

For this setup Hammer established the calibration curve [Hammer, 1980] linking the measured current and the amount of dissolved acidity in the ice. In our case the current is recorded by an x-y recorder, Watanabe Model 4410. Due to the slow response time of the recorder, short-time variations of the current are attenuated. With a sliding velocity of 10 cm sec^{-1}, the depth resolution is around 1 cm. To improve the resolution we tested a microprocessor controlled digital recording system in summer 1980 and 1981. For each mm moved with the electrodes, a reading of the digital voltmeter was recorded on an AIM 65 microcomputer. After each run the data were stored on a magnetic tape. The system, however, was not working satisfactorily. In summer 1980 there were problems with radar pulses from the Dye 3 station. For the last season the microcomputer was built into a faraday cage, but noise, most probably a 60 Hz oscillation, was superimposed on the recorded signal. The disadvantage of this low-cost system was that the recorded data could not be controlled in the field. Another approach was used during the field season 1982/1983 at the South Pole. The current was recorded as a frequency modulated signal on magnetic tape (Schwander, Ph.D. Thesis, University of Bern, 1984).

Physical Background of the ECM

If an electric field is applied on a piece of material by means of a pair of electrodes, a current starts to flow depending on the amount and mobility of the charge carriers and the amount of charge exchange at the electrodes. This situation is complicated for the ECM because the mobile charges in electrodes are electrons, while the current is believed to be transported by positive charged carriers in the ice. In ice the current with fixed electrodes is not stationary, in contrast to that in an ohmic resistor.

At the present stage, we can only give a qualitative explanation of this behavior. Von Hippel [1971] described qualitatively a mechanism that converts H_3O^+ and OH^- defects at the electrode-ice interface into electrons leaving or entering the electrodes. We suppose that the limiting factor of the initial current I_0 that we measure is the mobility and amount of positive charge carriers (H_3O^+ defects) in the ice. With time, the charge conversion at the electrodes becomes the limiting factor of the current. Because of the low mobility of the negative charge in the ice, the conversion at the positive electrodes starts to determine the current. At the same time, a space charge builds up. Figure 2 may help to visualize these processes. The initial current is a function of the amount of acids dissolved in the ice. Hammer [1980] indicates the following dependence.

$$(H^+) = 0.045 I_0^{1.73} \ \mu\text{equiv./kg} \ (T = -15°\text{C})$$

That such a well-defined correlation holds for acidity levels above 1 μequivalent/kg measured in the liquid phase, requires that only

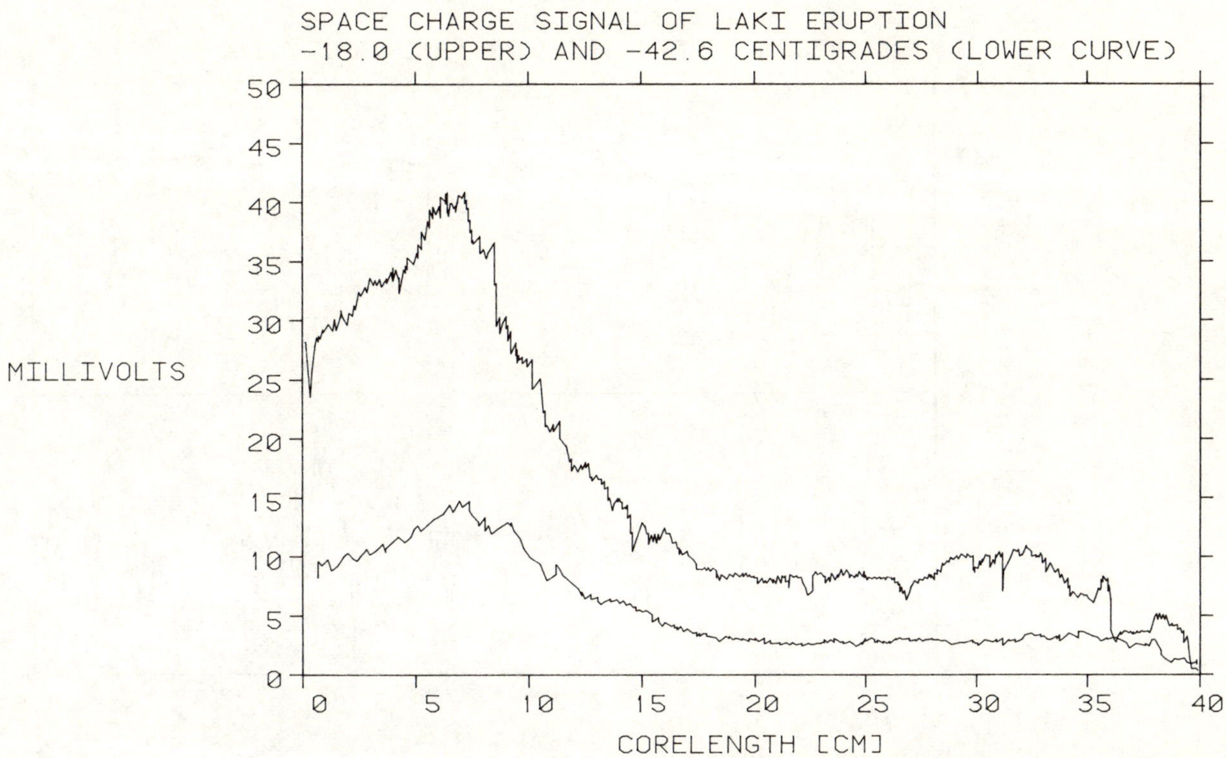

Fig. 3. ECM signal of Laki eruption measured at −18.0°C and −42.6°C.

strong acids such as HCl, H_2SO_4 and HNO_3 contribute substantially to the ECM signal and that they are incorporated into the ice matrix in ionic form. However, there are possible exceptions. When establishing a calibration curve between ECM current and acidity of the melted samples, one compares two different states; the solid state and the liquid state. While measuring the pH value, the ice has to be melted and during this step, chemical reactions between dissolved and undissolved impurities may occur. For example, sulphuric acid droplets attached to insoluble dust particles may lead to high acidity values in the liquid phase and low current values in the solid state. Soluble alkaline dust such as carbonates may lead to low acidity with relatively high current values. Such effects occur preferentially in ice with varying dust load such as cores from the Alps and ice from the Last Glaciation. The detailed comparison of the ECM with the chemical analysis will help to investigate whether such effects influence the Dye 3 ECM results.

If the acidity of the ice is below 0.1 μequivalent/kg, the current is at the detection limit of the used system in the field. For a given acidity, the current decreases with decreasing temperature. Statistical mechanics show that the mobility of an impurity in a solid depends exponentially on the temperature and therefore also on the measured current I_0. During the summer season 1979 we measured the largest ECM signal in the Dye 3 core, caused by the Laki eruption. Figure 3 shows this event measured at two different temperatures. Assuming an Arrhenius law for the temperature dependence $I = I_0 \exp(-E/kT)$, the activation energy was determined to be 0.23 ± 0.05 eV independent of the signal height. Therefore the activation energy is independent of the acidity concentration in the range of 1 to 20 μequivalent H^+/kg ice. This law does not hold for temperatures above −10°C where the surface current starts to be dominant. This is probably due to the fact that the liquid-like layer rapidly increases with temperature. The temperature dependence given above was used to normalize all results to the same temperature. Keeping this in mind, the ECM signals may be interpreted as "solid state pH value" for pH values below 6.5.

Discussion of the Measurements Made During the Deep Drill Seasons at Dye 3

The climatological interpretation of the ECM, together with other physical and chemical parameters measured in the field, will be given in another paper at this conference [Hammer et al., 1985]. We will focus our discussion on the potential of the information of the ECM.

In the Holocene ice there is good evidence that the calibration curve is valid [Hammer, 1980; Hammer et al., 1980]. Because the acidity content in the precipitation varies with the season (low values during winter, high values during summer), the ECM signal allows us to estimate the annual layer thickness until the first transition into Wisconsin ice at a depth of 1785 meters below surface. The mean ratio between summer and winter signal is around 2, corresponding to an acidity ratio of 3.3. Causes of these seasonal variations are discussed below. Higher ratios are found in years with a high percentage of melt layers. In general, higher signals were observed in melt layers, probably due to an uptake of acidic gases at the wet snow surface during summertime. Another explanation would be a deviation from the calibration curve, giving a higher current to the melt layers. This would be in opposition to a strong aging effect observed by Maccagnan et al. [1981] that depends upon the temperature being most pronounced close to the melting point.

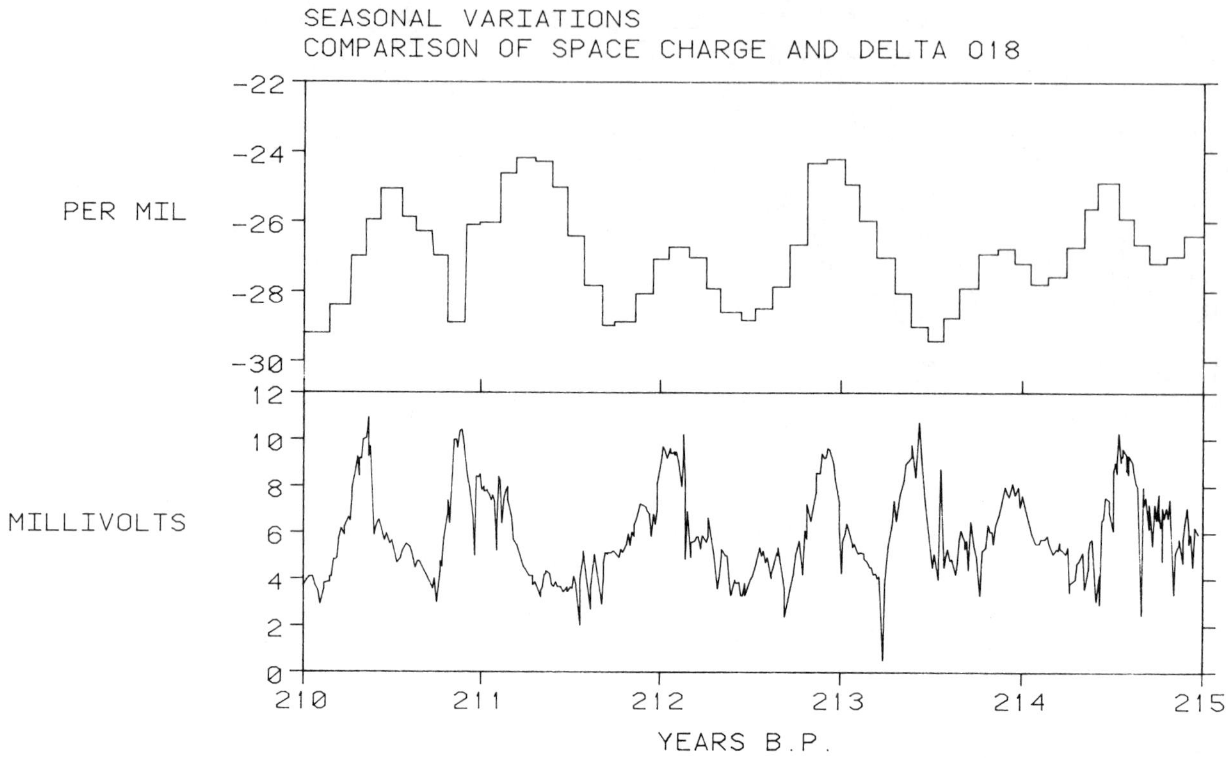

Fig. 4. Comparison of the seasonal features in the ECM signal and the $\delta^{18}O$ record.

There is a clear correlation of the ECM with the $\delta^{18}O$ measurements showing corresponding maxima in over 80% of all cases. For the Dye 3 core there are several possibilities of deviations, e.g., melt water percolated into the previous winter snow layer, volcanic influence, etc. These show the importance of cross-checking the ECM signal with other continuous records to count the years. As mentioned above, in years with a high percentage of melt layers higher signals were observed. It is therefore difficult to clearly distinguish low-volcanic signals from melt-layer signals. In such situations a quantitative comparison between stratigraphy, chemical analysis, $\delta^{18}O$ values, dust, and ECM may discriminate the two sources. Such an analysis has not yet been made. Figure 4 shows typical seasonal variations of the ECM signal at a depth of 865 m below surface, together with the $\delta^{18}O$ measurements. The seasonal variations are preserved until the first drastic transition into a colder climate at a depth of 1785.5 m below surface. Just before this transition, the annual layer thickness is approximately 1.5 cm. The difference between summer and winter level is still around 20% of the total signal. The amplitudes are lowered because of the diffusion of the impurities in the ice structure, and because in greater depth the two legs of the measuring electrode do not necessarily hit the same layer in the ice simultaneously.

Besides the volcanic signals, the most pronounced features in the ECM are the different level changes related to drastic climatic changes. Figure 5 shows the first and most astonishing transition from Holocene ice to Wisconsin ice, which occurred in a segment of 50 cm, corresponding to 30 years. Parallel to the decrease in the ECM is an increase in the content of insoluble dust particles observed by Hammer et al. [1985]. Part of this dust must be alkaline (carbonate dust) and is responsible for the low acidity content. Typical ''ice age'' ice from Greenland has a pH value of above 7. The signal recorded with the ECM in this ice is still 10 times higher than the signal recorded on a pure single crystal [Schwander et al., 1983]. This indicates that not all of the alkaline dust is dissolved in the ice. Remeasurements of such ''alkaline'' ice eight months later in the laboratory yield only 20% of the current measured in the field, and it is impossible to recognize the structure of the field data. We do not understand this result, but cannot exclude the possibility that during the storage time of the ice chemical reactions occurred in connection with the pressure relaxation.

The question arises whether a rapid stratigraphic change is necessarily caused by a fast climatic change. In an ice sheet, especially an ice sheet in motion, separation processes are in principle possible. Such a process for air bubbles is small and can be neglected [Stauffer, 1982]. A fast change of the atmospheric impurity concentrations and a changed precipitation rate (Herron, at this conference) are the most probable explanations, since a separation process leading to the observed rapid level change in the ECM and other parameters such as dust, $\delta^{18}O$, etc. is very unlikely.

Discussion of the Causes of the Regular Seasonal Variation of the ECM Signal

Herron [1982] has shown that nitric acid causes the regular seasonal variations of the ECM signal. The nitrate concentration in the ice can be used in the same way for dating the ice as can $\delta^{18}O$ for ice cores from the accumulation zone. The nitrate level is not elevated during active volcanic periods.

Nitric acid is a secondary product in the atmosphere and is formed in the troposphere by oxidation of NO_2 which in turn is formed by oxidation of NO. In the lower stratosphere the main source is oxidation of N_2O to NO with subsequent further oxidation to NO_2

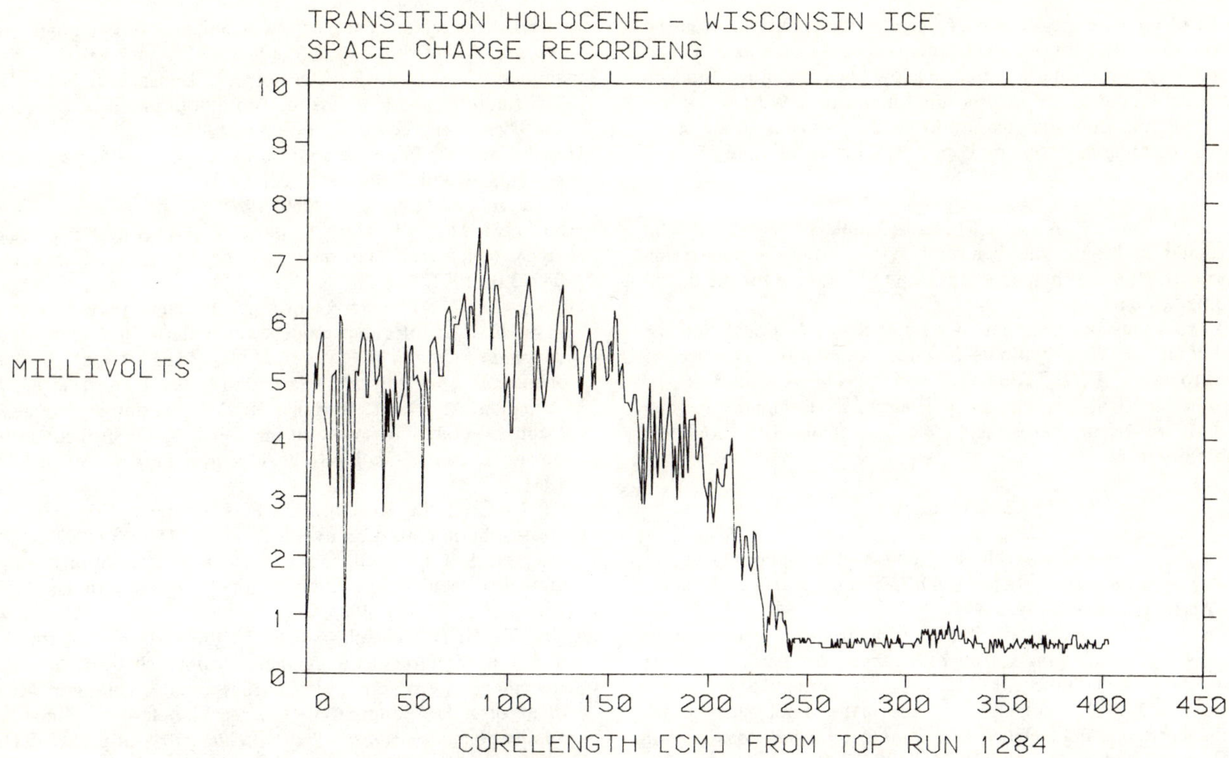

Fig. 5. ECM signal of the transition from Holocene to Wisconsin ice.

[Crutzen, 1979]. HNO_3 stands at the end of the oxidation chain of the oxidized N species in the troposphere and is removed by precipitation. Because of the high solubility in water, the mean residence time in the troposphere is assumed to be only 3 days [Hahn and Crutzen, 1982].

The production rate of HNO_3 in the troposphere depends upon the amount of NO_2 and the amount of OH radicals. The OH radical concentration in turn depends upon the amount of O_3, water vapor, and light intensity. Model calculations show an order of magnitude higher OH radical concentration in summer than in winter for high latitudes [Crutzen, 1983]. Therefore the photochemical production of HNO_3 must be higher in the summer if the NO_x ($NO + NO_2$) concentration is not the limiting factor.

Four different sources can be distinguished for the HNO_3 precursors in the troposphere under natural conditions [Hahn and Crutzen, 1982]: a) biomass burning, b) soils, c) lightning, d) downward transport from the stratosphere. The first two sources are continental sources and are predominantly located at low and middle latitudes. For Greenland, the large forest areas of northern North America are possible source regions. NO_x or HNO_3 from these sources would have to survive the transport through the atmospheric boundary layer, where the conditions for a fast oxidation are favorable (especially in summertime), and effective removal by precipitation takes place. Modelling of long range transport of nitrate and sulfate [Rhode, 1981] suggests that regional sources must be responsible for the nitrate in precipitation under today's conditions. We cannot expect that NO_x emitted into the troposphere by biomass burning will show a regular seasonal fluctuation over the last 10,000 years. It is, therefore, unlikely that NO_x produced by source a or b modulate directly the HNO_3 content in the Greenland ice.

NO_x production by lightning takes place partly in the free troposphere. One can speculate that the mean transfer distance for such NO_x is greater than for NO_x produced on the ground. The source strength of NO_x due to lightning is still a controversial matter, but could easily contribute 20% to the globally produced oxidized *N* species in the troposphere under natural conditions. Considering the fact that only a few percent of the atmospherically removed HNO_3 is deposited at latitudes greater than 65°, NO_x from lightning is a good candidate to modulate the nitrate concentration in ice because the amount of lightning shows a seasonal pattern [Turman et al., 1982].

Risbo et al. [1981] claimed that downward mixing of stratospheric air generates the higher nitrate deposition rate in the summer half year. Indeed in springtime the tropospheric-stratospheric exchange is enhanced as is shown by higher ^{10}Be, ^{7}Be, and ^{3}H levels in precipitation from this period of the year. Because of the assumed short atmospheric lifetime of HNO_3 we suppose a phaselag between the nitrate summer maxima and the $\delta^{18}O$ summer maxima which was not observed to date. Perhaps the resolution of the discrete NO_3^--concentration measurements in the ice was insufficient. A detailed comparison of ^{10}Be, ^{3}H, $\delta^{18}O$, and HNO_3 in a firn core covering the last 30 years could reveal whether a phaselag exists or not. Further, the integrated nitrate surplus in summer would be linked to the NO_x reservoir in the lower stratosphere which in turn can be influenced by intense stratospheric ionization events [Reid et al., 1978]. Assuming Risbo's hypothesis to be correct, the recorded nitrate concentration in an ice core, and therefore also the ECM, must reflect to a certain extent the stratospheric NO_x budget.

Conclusions

Taken alone, the ECM data are not more than a stratigraphical tool indicating seasonal variations and elevated levels due to volcanic eruptions. In the case of the different transitions during glacial time, ECM indicate the stepwise change of the acidity level in the ice. Only together with other continuous records such as $\delta^{18}O$ and dust and ion concentrations, do they allow correct dating of the ice and characterizing of the incorporated impurities. We are still far from calculating from the measured impurity concentrations in ice the corresponding atmospheric concentrations at the time of snow formation.

Acknowledgments. We would like to thank W. Bernhard who constructed and built the bandsaw system and the measuring platform. H. P. Moret helped to develop the software for the digital recording system.

This work has been supported by the Swiss National Science Foundation; the U.S. National Science Foundation, Division of Polar Programs, E. P. Todd, F. S. L. Williamson and R. L. Cameron; the U.S. Department of Energy, R. Dahlman; and the University of Bern, Switzerland. We also thank J. Tyndall for valuable discussions.

References

Crutzen, P. J., The role of NO and NO_2 in the chemistry of the troposphere and stratosphere, *Annual Reviews, Earth and Planetary Science Letters, 7,* 443–472, 1979.

Crutzen, P. J., The global distribution of hydroxyl in Atmospheric Chemistry, *Physical and Chemical Sciences Research,* Report 4, edited by E. D. Goldberg, p. 313, Springer Verlag, 1982.

Hahn, J., and P. J. Crutzen, The role of fixed nitrogen in the atmospheric photochemistry, *Philosophical Transactions, Royal Society of London, B*(296), 521–541, 1982.

Hammer, C. U., Acidity of polar ice cores in relation to absolute dating, past volcanism and radioechoes, *Journal of Glaciology, 25* (93), 359–371, 1980.

Hammer, C. U., H. B. Clausen, and W. Dansgaard, Greenland ice sheet evidence of post-glacial volcanism and its climatic impact, *Nature, 288* (5788), 230–235, 1980.

Hammer, C. U., H. B. Clausen, W. Dansgaard, A. Neftel, P. Kristindottir, and E. Johnson, Continuous impurity analysis along the Dye 3 deep core, this volume, 1985.

Herron, M. M., Impurity sources of F^-, Cl^-, NO_3^- and SO_4^{2-} in Greenland and Antarctic precipitation, *Journal of Geophysical Research, 87* (C4), 3052–4060, 1982.

Herron, M. M., S. L. Herron, and C. C. Langway Jr., The climatic signal of ice melt features in southern Greenland, *Nature, 293,* 389–391, 1981.

Hippel von, A., Transfer of protons through "pure" ice in single crystals II. Molecular models for polarisation and condition, *Journal of Chemical Physics, 54* (1), 145, 1971.

Hobbs, P. V., *Ice Physics,* Clarendon Press, Oxford, p. 83, 1974.

Maccagnan, M., J. M. Barnola, R. Delmas, and P. Duval, Static electrical conductivity as an indicator of the surface content of polar ice cores, *Geophysical Research Letters, 8* (9), 970–972, 1981.

Reid, G. C., J. R. McAfee, and P. J. Crutzen, Effects of intense stratospheric ionisation events, *Nature, 257,* 489–492, 1978.

Rhode, H., P. J. Crutzen, and A. Vauderpol, Formation of sulphuric and nitric acid in the atmosphere during long-range transport, *Tellus, 33,* 132–141, 1981.

Risbo, T., H. B. Clausen, and K. L. Rasmussen, Supernovae and nitrate in the Greenland ice sheet, *Nature, 294,* 637–639, 1981.

Schwander, J., A. Neftel, H. Oeschger, and B. Stauffer, Measurement of a DC-conductivity on ice samples for climatological applications, *Journal of Physical Chemistry,* in press, 1983.

Stauffer, B., Mechanismen des Lufteinschlusses in natürlichem Eis, *Zeitschrift für Gletscherkunde und Glazialgeologie, 16* (2), 1982.

Turman, B. N., and B. C. Edgar, Global lightning distributions at dawn and dusk, *Journal of Geophysical Research, 87* (C2), 1191–1206, 1982.

MECHANICAL PROPERTIES OF FRESH ICE CORE FROM DYE 3, GREENLAND

Hitoshi Shoji and Chester C. Langway, Jr.

State University of New York at Buffalo, Department of Geological Sciences, Ice Core Laboratory, Amherst, New York

Abstract. Mechanical property studies were carried out on fresh ice-core samples as they were recovered at Dye 3, Greenland. Experiments were conducted in uniaxial compression tests under constant crosshead speed, simpler shear tests under constant load, and indentation hardness tests with Vickers pyramidal indenter. Tests were made within a one-month period after each specimen was recovered to minimize the effect of volume relaxation. Experimental results from uniaxial compression and simple shear tests were analyzed in effective shear stress and effective strain rate space, taking into account the effect of high-pressure air bubbles included in each sample. The strain rate of Wisconsin ice is about one order of magnitude higher than that of randomly oriented laboratory ice at the same stress level, possibly enhanced by the effect of impurities in addition to that of fabrics. Indentation deformation was observed to be attributed to pileup, cracking, and creep of ice material around an indenter with c-axis orientation dependency.

Introduction

To understand the flow behavior of large polar ice masses, it is necessary to study the mechanical characteristics of natural glacier ice in the third (Z) dimension. This became possible with the advent of deep ice core drilling techniques that were developed for scientific purposes related to Greenland and Antarctic field-research programs. A new 2037 m deep ice core became available on August 10, 1981, with the successful completion of the National Science Foundation, Division of Polar Program (NSF/DPP) sponsored Greenland Ice Sheet Program (GISP). This deep ice core is probably the most carefully recovered and recorded ice core to reach bedrock from the polar ice sheets and will probably become the most comprehensively investigated to date. A completely integrated ice-core science study program was planned long before the core was obtained. One segment of the investigative plan was a detailed study of the mechanical properties as a function of core depth and time after recovery. It is known that ice cores undergo volume relaxation with time [Langway, 1958, 1962; Gow, 1971], which can affect the flow stress of the ice samples [Shoji and Higashi, 1978, 1979].

During the 1980 and 1981 field seasons of GISP at Dye 3 (65°12′N, 43°47′W) the mechanical property studies included uniaxial compression, simple shear, and indentation hardness tests on fresh core samples soon after each core sample was recovered (within a one-month period). The tests were made in a snow trench laboratory consisting of a recorder room (about −10°C) and a cave (about −16°C) as shown in Figure 1. This setup was made to keep the experimental temperature low and stable as possible.

The general physical characteristics of the Dye 3 ice core and the sampling depths for this work are shown in Figure 2. Below 600 m the ice became brittle with increasing depth and was badly fractured between 800 and 1200 m. The condition of the core thereafter progressively improved and below 1400 m was of excellent quality. Field microscopic observations revealed that air bubbles disappeared completely somewhere between 1500 and 1600 m. However, transparent air hydrate inclusions (Figure 3) appeared in the ice core between 1000 and 1300 m [Shoji and Langway, 1982b]. After the field work, follow-up observations were made at the Ice Core Laboratory and air hydrate inclusions were found in a core sample from 1200 m depth.

Details of the new core-drilling apparatus have been reported by Gundestrup and Johnsen [1982]. The electromagnetic coring device was retrieved up the fluid-filled borehole with an average speed of 1 m/s after each core cutting. This corresponds to a decreasing rate of hydrostatic pressure of about 5 bar/min for each core length recovered. The average borehole temperature was −20°C from near the surface to approximately the 900 m depth and −12°C at the bottom of the 2037 m hole. The borehole was slightly inclined and deviated 5.7° from vertical at the bottom.

Experimental Procedure

Specimen Preparation

Eight ice-core samples were selected over the vertical core profile apart from volcanic impurity layers detected by continuous conductivity measurement (Hammer, personal communication) as follows: 235, 504, 708, and 896 m depths from the 1980 ice core, and 939, 994, 1293, and 1814 m depths from the 1981 ice core. Sampling from the fractured zone (800 m to 1200 m) was made at a depth where unfractured core portion was long enough (~50 cm). Vertical slab sections (3.5 cm thick) were cut from each core sample as shown in Figure 4. Specimens for each test were rough cut with a band saw and finished with a surface microtome. All the specimens were free of cracks before each test was started.

Specimens for the uniaxial-compression tests were prepared with their uniaxial stress axis inclined 45° from the core axis, i.e., maximum resolved shear-stress plane parallel to the horizontal plane of the ice sheet. (Uniaxial stress is applied along the long direction of each specimen.) This preparation method makes a stress field under a uniaxial compression close to that of simple shear as related to resolved shear stress field in a specimen. Specimen dimensions and test data are given in Table 1.

Specimens for the simple-shear test were oriented to have the applied shear-stress plane parallel to the horizontal plane of the ice sheet. (The applied shear-stress direction was parallel to the 32 mm length direction.) Specimen dimensions and test data are given in Table 2. Silicone oil was applied to the specimen surfaces to minimize sublimation during testing.

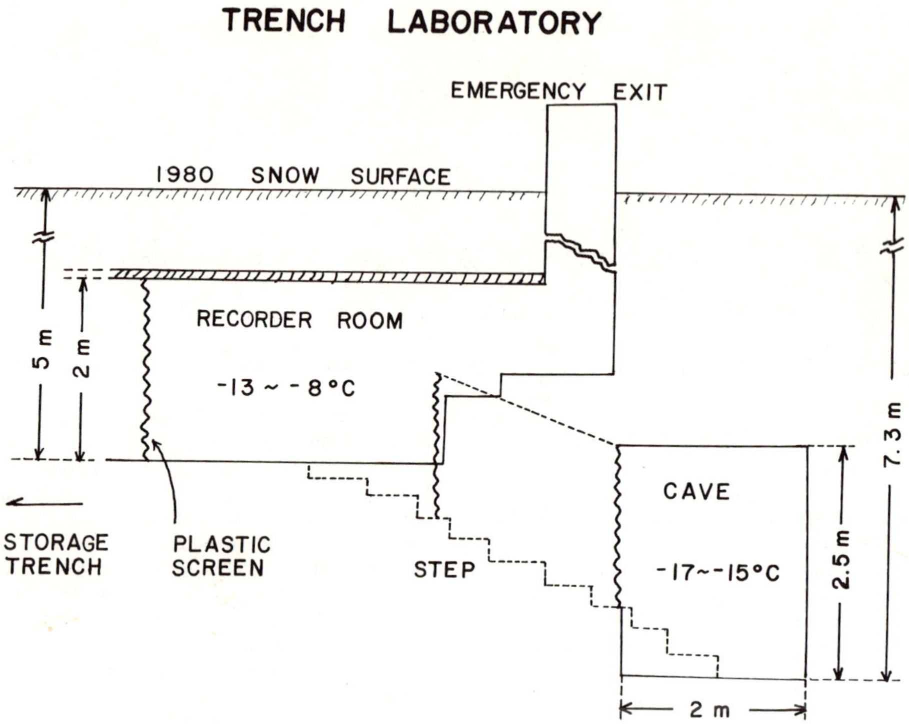

Fig. 1. Trench laboratories for mechanical property studies.

Indentation-hardness specimens were prepared from each depth core sample with horizontal and vertical orientation. Specimens (2 × 3 × 6 cm) were fastened to a glass plate by using a "tinker-dam" adhering technique [Shoji and Langway, 1982b] with a film of silicone oil between the specimen and glass plate. This freeze-on method minimizes thermal distortion and eliminates the problem of light scattering by air bubbles between the specimen and glass plate. The top surface was allowed to sublimate to obtain a planar surface.

Uniaxial Compression Test

Uniaxial-compression tests under constant crosshead speed were carried out with the Instron Model 1131 testing instrument which was especially modified for low-speed tests. One component of the apparatus consisting of the loading frame and driving unit was installed in the cave while the other component consisting of the controlling and load-recording unit was located in the recorder room

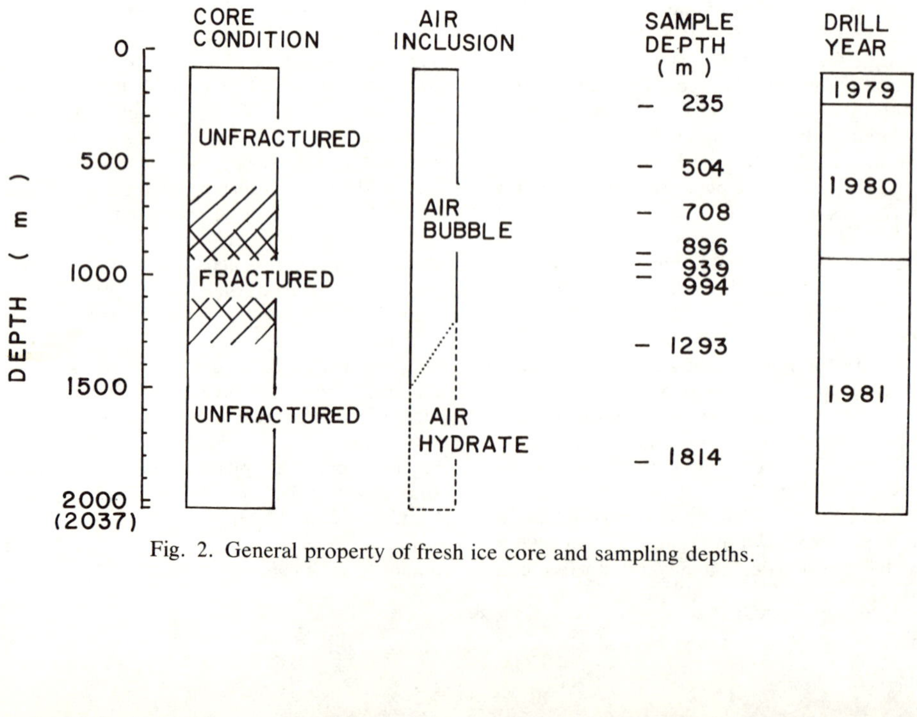

Fig. 2. General property of fresh ice core and sampling depths.

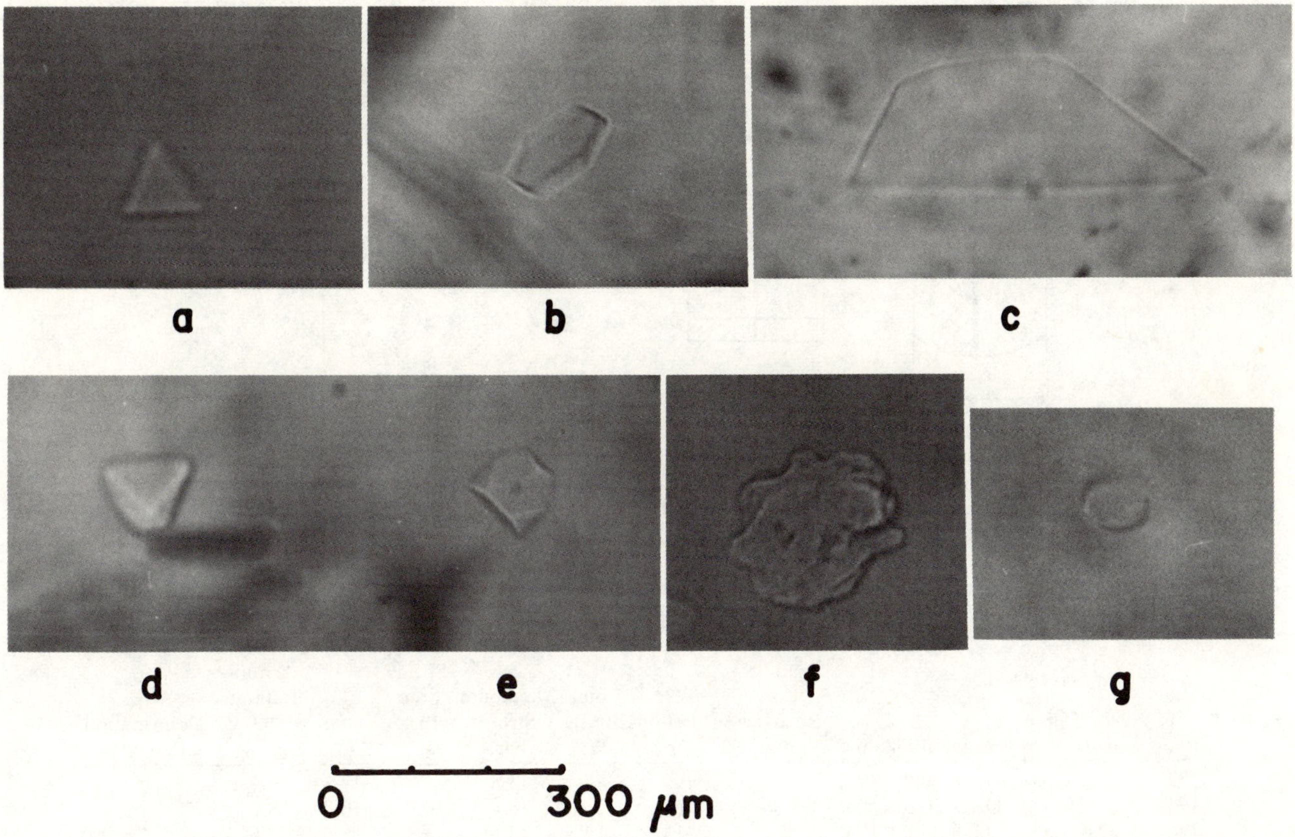

Fig. 3. Shape of air hydrate inclusion: triangular platelet, a; hexagonal platelet, b; polygonal platelet, c; tetrahedron, d; polyhedron, e; graupel-like, f; spheroid, g. [Shoji and Langway, 1982b].

shown in Figure 1. A displacement detector (LVDT, Pickering) was attached to the loading frame to measure the distance between the compression plates (equal to the specimen length). A copper-constantan thermocouple was used to measure the ambient temperature of the specimen.

Simple Shear Test

The simple-shear creep testing apparatus was especially designed and made at the Ice Core Laboratory. The specimen was frozen onto metal plates at both side faces. One of the plates was fixed on the apparatus frame and the other was allowed to slide vertically with linear motion bearing on vertical rods driven by a controlled weight. The vertical displacement of the sliding plate was recorded with the LVDT detector. The apparatus was contained in an insulation box installed in the cave with connecting wires to the recorders. Temperature was recorded with copper-constantan thermocouples.

Indentation Hardness Test

A Leitz-Miniload apparatus was used in the recorder room shown in Figure 1. A Vickers diamond indenter was used with a 55 g load. A total time of descent and indentation of 30 sec was adopted throughout the experiments. Temperature was recorded with a copper-constantan thermocouple. Optical microscopic observations were simultaneously carried out for precise measurement and recording impressions.

Experimental Results

The experimental results for the uniaxial compression tests are given in Table 1. Instrumental strain-rate range adopted for the tests was between 10^{-6} and $10^{-8}s^{-1}$. The very low fracture stress limited the number of data points for nonfracture tests. Fracture strength was 21.3 bar at 235 m (Specimen No. 4) and generally decreased with depth to 5.8 bar at 1293 m (Specimen No. 28). At a depth of 1814 m, the fracture strength is greater than 12 bar. The variation in fracture strength correlates well with the original quality of the core with depth. Two types of stress-strain curves were obtained as reported by Shoji [1978]. One type shows a yield drop related to crack generation in the specimen under high-strain rate, which is not the case of ice-sheet flow. The other is a stress-saturation type curve with no crack generation under low-strain rate. In some cases (Specimen Nos. 3 and 9) a few 2–3 mm diameter cracks were generated during the testing, but the stress-strain curve was a stress-saturation type (marked by an asterisk in Table 1). Temperature control was better than ± 0.2°C for each test but varied from one experiment to another due to the ambient weather changes. Temperature corrections were normalized at −16.0°C for each test by using an activation energy of 78.8 kJ/mol obtained with laboratory ice by Barnes et al. [1971]. Prior to the tests on ice-core specimens,

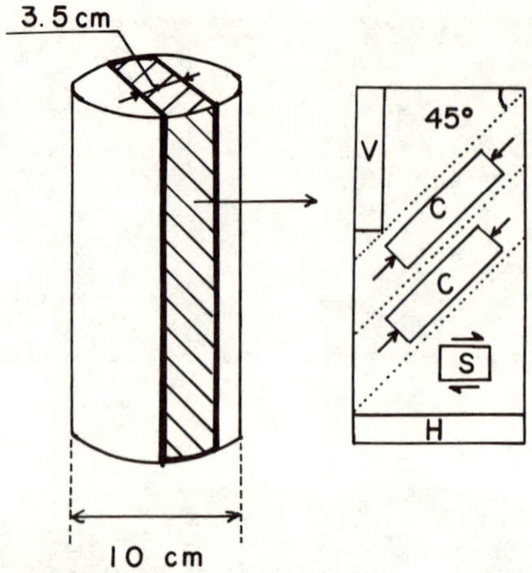

SAMPLE DEPTH (m)	GRAIN SIZE (mm)	C-AXIS VERTICALITY
235	2.4	FIRST QUARTILE φ(1/4)
504	4.5	
708	4.6	
896	2.2	20-30°
939	5.9	
994	4.1	
1293	5.6	
1814	1.2	10° (SINGLE MAX.)

Fig. 4. Specimen preparation and average value of grain size of each core sample. Fabric data is from Herron et al. [1982]. The first quartile of c-axis verticality is used by Russell-Head and Budd [1979] and measured from the orientation fabrics. For random fabric ice the first quartile would be 41.4°. With preferred orientation fabric ice, the value becomes smaller to about 10° for a single maximum fabric ice.

TABLE 1. Uniaxial Compression Data On Fresh Ice Core

Year of Rec.	Spec. No.	Depth m	Specimen Size mm	Onset Time after Core Recovery day	Test Duration hr	Total Strain %	Maximum Stress bar.	Strain Rate 10^{-7} s^{-1}	Temp °C	Corrected Strain Rate at −16.0°C 10^{-7} s^{-1}	Deformation Behavior
1980	1	235	80.1 x 20.0 x 19.7	3	106	4.0	15.8	1.0	−17.3	1.3	
1980	2	235	80.1 x 20.0 x 20.0	20	62	2.1	14.4	0.93	−16.8	1.0	
1980	3	235	79.8 x 20.0 x 20.0	22	16	1.9	23.5	3.3	−16.6	3.6	*
1980	4	235	80.7 x 20.1 x 20.1	23	4	1.4	21.3	14.	−16.6	—	Yield/Fract.
1980	5	235	79.8 x 20.0 x 19.9	23	10	4.1	16.1	11.	−16.2	12.	Yielding
1980	6	235	79.6 x 19.9 x 19.9	25	33	2.3	18.7	1.9	−16.6	2.1	
1980	7	235	79.9 x 20.0 x 19.9	27	92	1.7	11.3	0.51	−16.3	0.54	
1980	8	504	82.6 x 26.1 x 24.2	8	12	1.4	4.6	3.1	−16.1	3.2	Yielding
1980	9	504	83.1 x 25.4 x 24.0	9	23	1.3	13.3	1.6	−16.0	1.6	*
1980	10	504	82.5 x 25.6 x 24.1	10	9	1.3	14.0	4.4	−15.8	—	Yield/Fract.
1980	11	504	82.6 x 26.8 x 26.3	12	109	2.0	6.9	0.50	−15.8	0.49	
1980	12	504	82.7 x 26.0 x 24.1	17	111	2.4	7.5	0.61	−15.8	0.59	
1980	13	504	80.6 x 26.7 x 25.9	22	17	0.6	7.3	0.91	−15.9	—	Fracture
1980	14	504	82.4 x 26.5 x 26.4	23	68	2.5	13.0	1.0	−15.9	1.0	
1980	15	708	83.0 x 28.0 x 26.5	6	46	2.0	10.1	1.2	−15.5	—	Fracture
1980	16	708	81.3 x 26.7 x 26.5	8	38	0.7	9.8	0.49	−15.8	—	Fracture
1980	17	708	83.2 x 28.1 x 25.8	12	100	1.9	8.9	0.52	−15.0	0.45	
1980	18	896	83.3 x 29.3 x 28.3	5	7	1.1	19.3	4.1	−14.8	—	Fracture
1980	19	896	83.5 x 29.1 x 28.1	5	7	0.9	14.1	3.8	−14.7	—	Fracture
1981	20	939	81.5 x 27.0 x 25.6	9	71	2.8	10.3	1.1	−16.8	1.3	
1981	21	939	79.8 x 25.5 x 22.2	13	70	1.6	9.0	0.64	−16.4	0.68	
1981	22	939	90.7 x 25.7 x 22.2	29	5	0.6	10.5	3.7	−16.2	—	Fracture
1981	23	994	89.9 x 26.1 x 24.1	12	161	2.5	8.0	0.44	−16.5	0.47	
1981	24	994	86.8 x 25.4 x 21.7	21	39	3.4	13.3	2.4	−16.4	2.5	Yielding
1981	25	994	72.2 x 22.0 x 20.1	23	8	0.5	11.0	2.0	−16.4	—	Fracture
1981	26	994	75.3 x 22.4 x 21.8	24	135	3.8	5.2	0.77	−16.8	0.87	
1981	27	1293	81.3 x 25.9 x 25.9	7	111	3.3	8.1	0.82	−16.9	0.93	
1981	28	1293	80.4 x 24.9 x 24.5	13	25	0.6	5.8	0.69	−16.9	—	Fracture
1981	29	1293	84.0 x 24.9 x 24.5	16	48	1.7	5.3	0.98	−16.6	1.1	Yielding
1981	30	1293	82.5 x 25.0 x 24.8	28	9	0.7	6.2	2.0	−15.8	—	Fracture
1981	31	1293	84.3 x 25.5 x 25.2	29	209	8.0	5.7	1.1	−15.6	1.0	
1981	32	1814	87.1 x 30.0 x 24.8	7	84	5.9	6.6	2.0	−15.7	1.9	
1981	33	1814	85.8 x 28.7 x 24.4	11	20	15.8	12.2	21.	−16.0	—	Buckling
1981	34	1814	85.0 x 29.3 x 24.5	12	29	10.7	10.5	10.	−16.1	11.	
1981	35	1814	82.6 x 27.5 x 23.9	13	22	4.1	9.6	5.0	−16.3	5.3	

TABLE 2. Simple Shear Data On Fresh Ice Core

Year of Rec.	Spec. No.	Depth m	Specimen Size mm	Onset Time after Core Recovery day	Test Duration day	Total Strain %	Stress bar.	Minimum Strain Rate 10^{-9} s^{-1}	Temp °C	Corrected Strain Rate at −16.0°C 10^{-9} s^{-1}
1980	1	235	32.0 x 22.3 x 21.4	3	30	1.3	1.8	2.2	−17	2.5
1980	2	504	32.0 x 22.3 x 21.4	11	14	2.0	1.8	8.7	−16	8.7
1980	3	708	31.5 x 21.3 x 21.4	2	15	1.3	1.9	3.2	−16	3.2
1981	4	939	32.1 x 22.4 x 21.4	10	20	3.3	1.2	8.3	−17	9.6
1981	5	994	31.3 x 22.3 x 21.4	24	28	4.9	1.5	25.	−18	33.
1981	6	1293	31.9 x 22.5 x 21.4	29	10	2.3	1.5	19.	−17	22.
1981	7	1814	32.2 x 22.5 x 21.4	7	8	4.9	1.5	35.	−17	41.

reference specimens from ice layers near the surface with random orientation fabrics were examined each year. The results agreed quite well with the above results on laboratory ice.

The simple shear test results are shown in Table 2. Temperature control was held within ± 1°C for each experiment. Each test was continued until the total shear strain was more than 1.3% for the minimum strain rate stage. Temperature corrections were similar to those made for the uniaxial compression tests.

The results of the indentation hardness test data are given in Table 3. Indentation traverses were made on each specimen with a spacing of about 1mm between impressions. Impressions within 50 μm from a grain boundary or air bubble on the specimen surface were not adopted for data analyses. Characteristic images around each impression were observed by optical microscope and were found to be strongly related to the basal plane orientation of the ice grain impressed. Four typical examples are shown in Figure 5. Frequency of appearance of these types is quite different between horizontal- and vertical-section specimens. With horizontal-section specimens, type A and B are very rare. The sum of the frequency of their appearance is within 25% of the total indentations on each depth specimen. However, with vertical section specimens, type A and B impressions amount to over 75% of the total on each depth specimen. All tests were performed within 3 weeks after each specimen was recovered from the borehole.

Discussion of Results

Uniaxial Fracture Stress

One of the remarkable results of the fresh-ice core under unconfined pressure conditions is its quite low fracture stress down to about 6 bar with 1293 m depth specimens as shown in Table 1. This brittleness of the ice core certainly corresponds to the core-quality profile shown in Figure 2. Since the indentation-hardness number obtained shows no significant difference from those of artificial ice, as discussed later, this decrease in fracture stress should be attributed to macroscopic-structure change of the ice core such as grain size and high-pressure air bubbles rather than to the ice structure itself.

Muguruma [1969] found a grain-size effect on yield (fracture) stress, σ_f, in uniaxial compression with columnar-grained polycrystalline ice (Hall-Petch relationship). He showed that σ_f is given by

$$\sigma_f = 7.5 + 31\ d^{-1/2} \tag{1}$$

where d is the average grain diameter in mm and σ_f is expressed in bar. This equation can be derived from the assumption that the work done on the number of dislocations piled up is transferred to the internal surface energy of a cleavage crack.

Stress field around a high-pressure air bubble is given by Nye [1953, 1957] as follows:

$$\tau_0 = \frac{3^{1/2}}{2n}\left(\frac{a}{r}\right)^{3/n} \Delta P \tag{2}$$

where τ_0 and ΔP are effective shear-stress and hydrostatic-pressure difference between that in an air bubble and ambient pressure, respectively. a is the bubble radius and r is the distance from the center of the bubble. n is the flow-law parameter given as 2.7 by Shoji and Langway [1983]. For simplicity, we take this value as 3. Averaging the stress, τ_0, over a plane of a radius, r^* (r^*: half value of the mean bubble-to-bubble distance), we obtain, provided that $r^* \gg a$,

TABLE 3. Vickers Identation Hardness Number (VHN) On Fresh Ice Core

	on Horizontal Section				on Vertical Section			
Depth m	VHN kg/mm^2	Standard Deviation	No. of Indentations	T °C	VHN kg/mm^2	Standard Deviation	No. of Indentations	T °C
235	5.0	0.6	126	−13	5.3	1.0	116	−12
504	5.1	0.7	126	−11	5.2	0.8	108	−11
708	5.0	0.8	164	−11	4.4	0.7	119	−12
896	4.9	0.6	80	−11	4.7	0.5	94	−11
939	5.3	0.7	104	−13	—			
994	5.4	0.7	107	−12	6.0	1.7	115	−14
1293	4.9	0.4	176	−11	5.6	0.7	175	−11
1814	—				5.0	0.7	105	−8

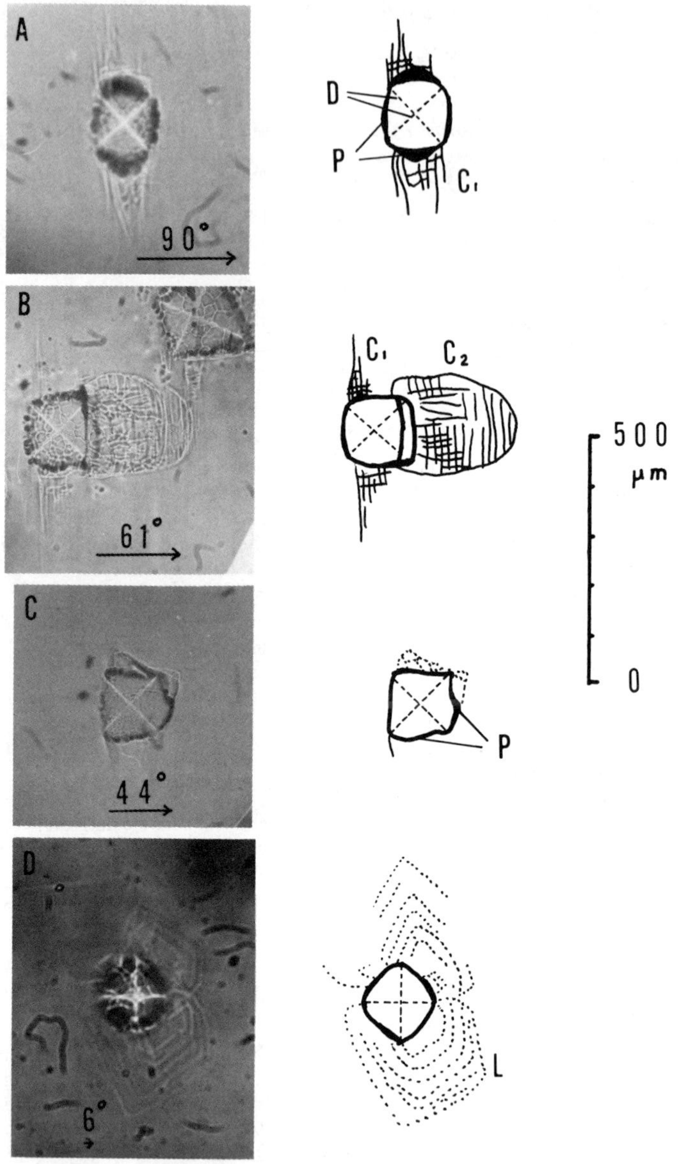

Fig. 5. Indentation pattern around impression. Numerals on each arrow indicate angles between c-axis and the normal to specimen surface. The arrows are projections of c-axis orientation on the specimen surface. Indentation diagonal (D), pileup around impression (P), basal cracks with non-basal crack segments (C_1), tongue-like cracked zone (C_2), multiple loops of sub-boundary grooves (L).

$$\bar{\tau}_1 = 3^{-1/2}\left(\frac{a}{r^*}\right)\Delta P \quad (3)$$

Since the stress field around a bubble is radial, this mean value along a plane, $\bar{\tau}_1$, gives a maximum stress along a slip direction in a grain to be added to the applied stress for dislocation pileup and subsequent onset of cleavage crack. Therefore, fracture stress in effective shear stress, τ_f, may be given by

$$\tau_f = \tau_a + \bar{\tau}_1 \quad (4)$$

where τ_a is effective shear stress from applied uniaxial compression, σ_a, ($\tau_a = 3^{1/2}\,\sigma_a$). Air-bubble density and radius measured with fresh ice core are given in Table 4 which includes $\bar{\tau}_1$ values calculated. The effective shear stresses for fracture (including data of yielding and generation of a crack), τ_f calculated are shown in Figure 6 at each grain size level. The solid line by Muguruma fits quite well with data obtained. This agreement suggests that fracture stress *in situ* in the ice sheet is the same as that of artificial ice. However, another effect of hydrostatic pressure on fracture stress which might be taken into account was reported by Jones [1982] with artificial polycrystalline ice. According to his experiments, fracture (yielding) stress increased with the increasing hydrostatic pressure up to about 200 bar at high strain-rate tests.

TABLE 4. Air Bubble Density and Radius

Specimen Depth m	Measured Date After Core Recovery day	Bubble Density N cm^{-3}	Bubble Radius a μm	Half Bubble to Bubble Distance $r^* = (N^{-1/3})/2$ μm	Maximum Stress $\bar{\tau}_1$ bar	Mean Stress $\bar{\tau}_0$ bar
235	37	320	157	730	2.6	1.1
504	14	200	132	855	4.0	1.7
708	1	300	103	745	5.1	2.2
896	4	290	95	755	5.9	2.6
939	15	330	105	725	7.1	3.1
994	13	330	105	725	7.5	3.2
1293†	—	300	90	745	8.1	3.5
1814‡‡	3	1000	10	500	1.9	0.8

† No measurement was made on this depth specimen. N and a were estimated from the other depths data.
‡‡ No preexisting air bubbles were detected immediately after core recovery. In 3 days, air bubbles appeared on air hydrate inclusions (a ~5μm) and grain boundaries (a ~1μm) in cloudy band zone. Density of air hydrate inclusion is 1870 cm^{-3} at this depth both in cloudy and clear zones. The above value of N and a were estimated. ΔP = 96 bar (dissociation pressure of air hydrate at −16°C by Miller [1969].

Glacier Flow

Our experimental results show a wide range of ice-core strength data for non-fractured specimens as shown in Table 1. The above discussion with analytical treatment suggests that flow behavior in low-strain rate regions is also affected by the stress field around a high-pressure air bubble. However, creep properties at low-strain rates should not depend on a localized stress field as in the case of fracture but depend on a more averaged stress field in each specimen. According to Nye [1953], effective strain rate, $\dot{\varepsilon}$, and effective shear stress, τ, are expressed as follows:

$$\dot{\varepsilon} = B\tau^n \quad (5)$$

where B and n are constants obtained from flow-law parameters. The above equation can be interpreted through energy consideration. That is, the strain rate depends on stored energy, $\tau^2/(2G)$, where G is an elastic constant, as well as stress deviator. Therefore, we reasonably assume that the effect of a high-pressure air bubble appears through effective shear stress, τ, as follows:

$$\tau^2 = \tau_a^{\,2} + \bar{\tau}_0^{\,2} \quad (6)$$

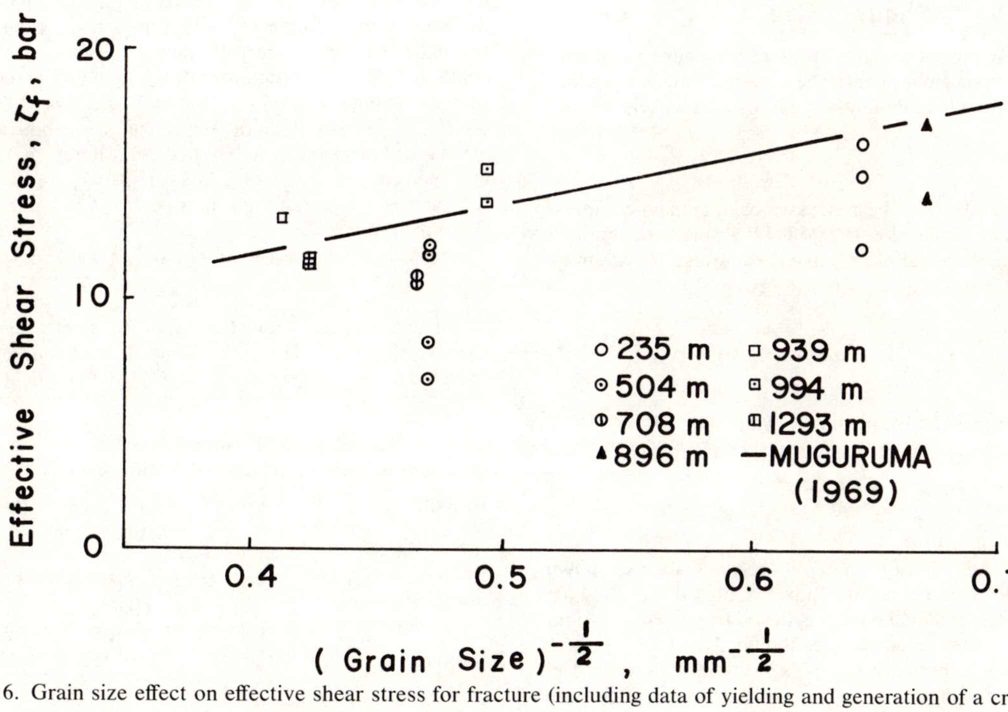

Fig. 6. Grain size effect on effective shear stress for fracture (including data of yielding and generation of a crack).

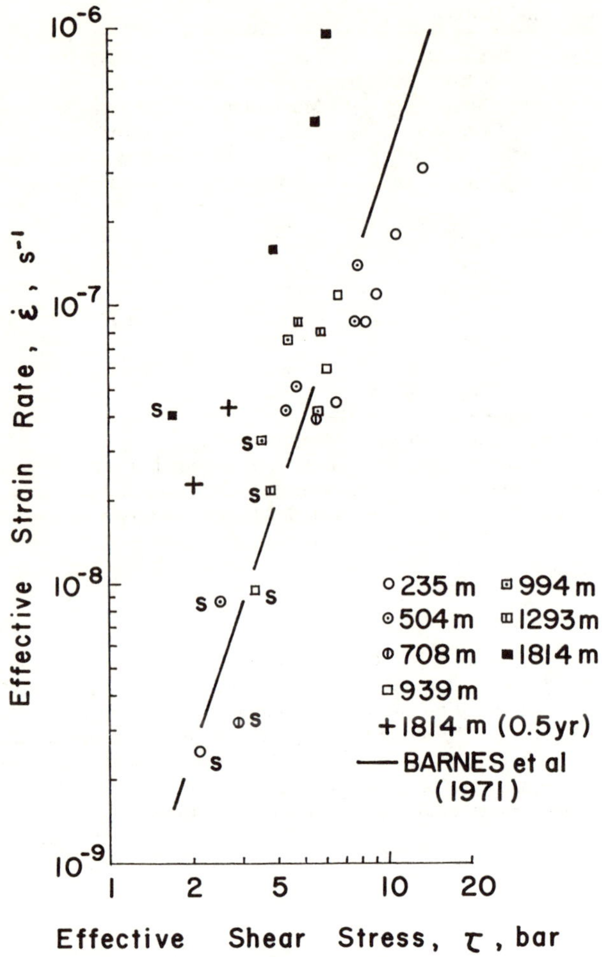

Fig. 7. Logarithmic plot of effective strain rate vs. effective shear stress. Data converted from simple shear results are marked S. Data on randomly oriented laboratory ice is also shown [Barnes et al., 1971].

where τ_a and $\bar{\tau}_0$ are effective shear stresses from applied compressive stress, σ_a, and from the stress field around the bubble, respectively. The mean value of effective shear stress, $\bar{\tau}_0$, around a bubble is given by

$$\bar{\tau}_0^2 = \int_a^{r^*} \pi r^2 \, \tau_0^2 \, dr/\{4\pi(r^*)^3/3\} \tag{7}$$

From equations (2) and (7), we obtain, provided that $r^* \gg a$,

$$\bar{\tau}_0 = \frac{\Delta P}{4}\left(\frac{a}{r^*}\right) \tag{8}$$

The values of $\bar{\tau}_0$ calculated for each depth specimen are also shown in Table 4. The corrected results are shown in Figure 7 as a log-log plot of effective-strain rate and effective-shear stress. The contribution of volume expansion of air bubbles to effective-strain rate was neglected. In Figure 7 there are two points obtained from 1814 m depth specimens of which tests were made 0.5 year after core recovery. With the above results, no stress correction was made for the high-pressure bubble effect. The results from simple shear tests are denoted by S, which agree well with uniaxial compression test data within a scatter of data points, except for the result from 1814 m depth specimen. Since the 1814 m depth specimen is of a single maximum fabric, the deformation is quite anisotropic. Therefore, Nye's treatment for stress conversion is invalid and the significant portion of specimen volume near compressive platens may be left as non-deformed parts and subsequently does not contribute to total strain of the specimen [Jones and Brunet, 1978]. These effects may apparently decrease the strain-rate values obtained from uniaxial tests relative to those from simple shear test. The solid line shows the results of artificial polycrystalline ice by Barnes et al. [1971]. Data points show that the stress exponent (slope in Figure 7) is equal to or close to 3.

Fabric studies on the same fresh ice core were carried out by Herron et al. [1982]. Specimens from depths of 504, 939, 994 and 1293 m have fabrics of random to vertical cluster and their measured enhancement factor of 1 to 2 agrees with those from the fabric enhancement factor diagram [Russell-Head and Budd, 1979]. For the 708 m depth specimens of girdle-like fabric, the factor measured was less than 1. Lile [1978] obtained an enhancement factor of 0.83 on girdle fabric specimens in uniaxial compression with the stress axis perpendicular to the pole direction of the girdle pattern, which is very close to the above value. However, for specimens from 235 m depth the enhancement factor measured is between 0.5 and 0.8, although the specimens have quite random orientation fabric. This low enhancement factor might be related to the even lower value (0.4~0.1) obtained from borehole closure measurements at shallow depths of ice sheets [Paterson, 1977].

For specimens from 1814 m depth (single maximum fabric), the enhancement factor measured is about 8 to 12 which is obviously 2 to 3 times higher than the maximum value of 4 for the fabric enhancement factor [Russell-Head and Budd, 1979]. Nakamura and Jones [1973] studied impurity effects on yield with a single crystal of ice. According to their experiments, 1.3 ppm HCl doping decreased the yield stress up to one-half of that of pure ice. This corresponds to the enhancement factor of 8 when the stress exponent, n, is 3. The content of Cl^- measured with the late Wisconsin ice (below 1786 m depth at Dye 3) is in the order of 0.1 ppm and about 5 times higher than the Holocene level [Herron and Langway, 1982]. The content of SO_4^{2-} is also high in Wisconsin ice, though the effect on mechanical property is not studied yet. Therefore it is quite possible that the remnant factor of 2 to 3 for the strain-rate enhancement is caused by the effect of impurities such as Cl^- and SO_4^{2-} for the late Wisconsin ice.

Indentation hardness results listed in Table 3 shows an almost

TABLE 5. Indentation Hardness Number, kg/mm^2 on Single Crystal of Ice at −11°C. The Long Axis of the Knoop Indenter is Inclined 45° (C_{45}), Perpendicular ($C_\perp$) and Parallel ($C_\parallel$) to the c-axis

Indenter	On Basal Plane	On Prism Plane			Specimen	Reference
Brinell	6.3	6.2			Glacier Ice	Butkovich [1954]
Knoop		C_{45}	$C_\perp$	$C_\parallel$	Artificial Ice	Offenbacher and Roselman [1971]
		4.5	5.2	6.8		
	4.0		3.3	4.1	Artificial and Glacier Ice	Ackley [1973]
	4.3	5.6	5.8	7.5	D_2O Ice	Offenbacher [1977]
Vickers		5.2 ± 0.8			Fresh Ice Core	This Work

Fig. 8. Air bubble shape one day after core recovery from a depth of 708 m. The plane of crack ring is measured to be parallel to basal plane by hoar-crystal method [Shoji and Langway, 1982a], which refers to the hoar-crystal direction grown on a specimen surface. The bubble diameter along basal plane is 300 μm.

constant level of hardness over the profile sampled. The mean value of 5.2 ± 0.8 kg/mm^2 obtained in this work is slightly less than that in the Brinell hardness experiments and within the scatter of Knoop hardness data as shown in Table 5. These variations may be caused by the differences of stress fields among each indenter [Tolansky and Nickols, 1949; Thibault and Nyquist, 1947]. It has been revealed that two types of slip systems are activated by Knoop indentation tests: basal- and non-basal systems [Offenbacher, 1971]. The effective resolved shear stress, τ_e, is calculated from the tensile forces parallel to the line of steepest slope on the indenter's individual facets and taking into account the rotational constraint of slip planes [Brookes et al., 1971]. It can be seen that subsurface basal cracks with nonbasal-crack segments appear only when the resolved shear stress on the basal plane is close to zero as C_1 and C_2 in Types A and B of Figure 5. When basal slip is permitted, we cannot expect stress concentrations high enough to nucleate and propagate basal cracks. Since ideal fracture strength does not depend on temperature, this may explain less temperature dependency of Brinell hardness number measured on prism plane than on basal plane [Butkovich, 1954].

From the apparent similarity of the stress fields around the indentation imprints as well as around air bubbles with high internal pressure, we are able to understand more fully the bubble expansion process. An example of an air bubble in a core sample from a depth of 708 m one day after the sample recovery is shown in Figure 8. Crack-ring images around the slightly elongated bubble are to be seen. The plane of these rings are measured to be parallel to the basal plane by the hoar-crystal method [Shoji and Langway, 1982]. This means that these crack rings are identical to the basal cracks around the indentation imprints. The storage temperature for this sample was about −16°C. Therefore we can expect that the bubble elongates along the basal plane slightly, as is seen in Figure 8, since ice is harder along the c-axis than along the basal plane at this temperature [Butkovich, 1954].

Summary

(a) Fracture stress depends on grain size (Hall-Petch relationship) and maximum effective shear stress, $\bar{\tau}_1$, caused by high-pressure air bubbles under unconfined pressure conditions.

(b) The enhancement factor in strain rate is interpreted as a coupling of the fabric factor and the impurity factor. The highest enhancement factor of about 10 was obtained with late Wisconsin ice.

(c) Fresh ice core hardness data showed almost the same hardness level as artificial ice data. Indentation deformation is attributed to pileup, cracking, and creep of ice around an indenter with c-axis orientation dependency.

Acknowledgments. The many discussions with K. Itagaki were invaluable. We wish also to thank W. Bernhard for technical help, and N. Tseng and W. Peterson for assistance in the field laboratory. This research work was supported by the National Science Foundation, Division of Polar Programs.

References

Ackley, S. F. Microhardness testing on ice single crystals, *Physics and chemistry of ice*, (paper presented at the Symposium held in Ottawa, Canada, August, 1972), edited by E. Whalley and others, 382–386, Royal Society of Canada, Ottawa, 1973.

Barnes, P., D. Tabor, and J. C. F. Walker, The friction and creep of polycrystalline ice. *Proceedings of the Royal Society of London, A. 324*, 127–155, 1971.

Brookes, C. A., J. B. O'Neill, and B.A.W. Redfern, Anisotropy in the hardness of single crystals, *Proceedings of the Royal Society of London, A. 322*, 73–88, 1971.

Butkovich, T. R. hardness of single ice crystals, *Research Report 9*, 12 pp., U.S. Army Corps of Engineers, Snow, Ice and Permafrost Establishment, Wilmette, Ill., 1954.

Gow, A. J. Relaxation of ice in deep drill cores from Antarctica, *Journal of Geophysical Research, 76*, 2533–2541, 1971.

Gundestrup, N. and S. J. Johnsen, A battery powered, instrumented deep ice core drill for liquid-filled holes (abstract), *EOS Transactions, American Geophysical Union, 63*, 297, 1982.

Herron, S., C. C. Langway, Jr., and K.A. Brugger, Ultrasonic velocities and crystalline anisotropy in the ice core from Dye 3, Greenland (abstract), *EOS Transactions, American Geophysical Union, 63*, 297, 1982.

Herron, M. M. and C. C. Langway, Jr., Chloride, nitrate, and sulfate in the Dye 3 and Camp Century, Greenland ice cores (abstract), *EOS Transactions, American Geophysical Union, 63*, 298, 1982.

Jones, S. J. The confined compressive strength of polycrystalline ice, *Journal of Glaciology, 28*, 171–177, 1982.

Jones, S. J. and J. G. Brunet, Deformation of ice single crystals close to the melting point, *Journal of Glaciology, 21*, 445–455, 1978.

Langway, C. C., Jr. Bubble pressures in Greenland glacier ice, Symposium of Chamonix, Physics of the Movement of Ice (Proceedings of the Chamonix Symposium, 1958), *IAHS-AISH Publication No. 47*, 336–349, 1958.

Langway, C. C., Jr. Some physical and chemical investigations of a 411 meter deep Greenland ice core and their relationship to accumulation, Symposium of Obergurgl, Variations of the Regime of Existing Glaciers (Proceedings of the Obergurgl Symposium, 1962), *IAHS-AISH Publication No. 58*, 189–198, 1962.

Lile, R. C. The effect of anisotropy on the creep of polycrystalline ice, *Journal of Glaciology, 21*, 475–483, 1978.

Miller, S. L. Clathrate hydrates of air in Antarctic ice, *Science, 165*, 489–490, 1969.

Muguruma, J. Effects of surface condition on the mechanical properties of ice crystals, *British Journal of Applied Physics (Journal of Physics, D), 2*, 1517–1525, 1969.

Nakamura, T. and S. J. Jones, Mechanical properties of impure ice crystals, *Physics and chemistry of ice*, (paper presented at Symposium held in Ottawa, Canada, August, 1972) edited by E. Whalley and others, 365–369, Royal Society of Canada, Ottawa, 1973.

Nye, J. F. The flow law of ice from measurements in glacier tunnels, laboratory experiments and the Jungfraufirn borehole experiment, *Proceedings of the Royal Society of London, Ser. A, 219*, 477–489, 1953.

Nye, J. F. The distribution of stress and velocity in glaciers and ice sheets, *Proceedings of the Royal Society of London, Ser. A., 239*, 113–133, 1957.

Offenbacher, E. L. and I. C. Roselman, Hardness anisotropy of single crystals of ice in nature, *Physical Sciences, 234*, 112–113, 1971.

Offenbacher, E. L. Hardness anisotropy of deuterium oxide single crystals, Symposium on Isotopes and Impurities in Snow and Ice (Proceedings of the Grenoble Symposium, 1975), *IAHS-AISH Publication No. 118*, 44–49, 1977.

Paterson, W. S. B. Secondary and tertiary creep of glacier ice as measured by borehole closure rates, *Reviews of Geophysics and Space Physics, 15*, 47–55, 1977.

Russell-Head, D. S. and W. F. Budd, Ice-sheet flow properties derived from borehole shear measurements combined with ice-core studies, *Journal of Glaciology, 24*, 117–130, 1979.

Shoji, H. and A. Higashi, X-ray diffraction topographic studies of Antarctic deep core ice. *Japanese Journal of Applied Physics, 17*, 993–1001, 1978.

Shoji, H. Stress-strain tests of ice core drilled at Mizuho Station, East Antarctica. *Memoirs of National Institute of Polar Research in Japan, Special Issue Number 10*, 95–101, 1978.

Shoji, H. and A. Higashi, Mechanical properties of Antarctic deep-core ice, *Journal of Glaciology, 24*, 487–489, 1979.

Shoji, H. and C. C. Langway, Jr., Hoar crystal method for ice fabrics, *Journal of Glaciology, 28*, 387–390, 1982a.

Shoji, H. and C. C. Langway, Jr., Air hydrate inclusions in fresh ice core, *Nature, 298*, 548–550, 1982b.

Shoji, H. and C. C. Langway, Jr., Volume relaxation of air inclusions in a fresh ice core, *Journal of Physical Chemistry, 87*, 4111–4114, 1983.

Thibault, N. W. and H. L. Nyquist, The measured Knoop hardness of hard substances and factors affecting its determination, *Transactions of the American Society for Metals, 38*, 271–330, 1947.

Tolansky, S. and D. G. Nickols, Interferometric examination of hardness test indentations, *Nature, 164*, 103–104, 1949.

BEDROCK TOPOGRAPHY OF THE GREENLAND ICE SHEET IN THE DYE 3 AREA

S. Overgaard[1]

Electromagnetics Institute, Technical University, Denmark

N. S. Gundestrup

Geophysical Isotope Laboratory, University of Copenhagen, Denmark

Abstract. The bedrock topography of a 1400 square kilometer area around the Dye 3 radar station on the Greenland Ice Sheet is obtained from radio-echo soundings. The subglacial landscape is very mountainous with large changes in bedrock relief. The compiled map of the bedrock shows that the deep ice core from Dye 3 is on the eastern slope of a valley that trends northwest-southeast. In addition there are large undulations upstream of the drill site. Measurements were made by both an airborne 10 kw sounder, and a 1 kw lightweight surface sounder. Corrections for navigational errors were made to the airborne measurements when data from both were combined to produce topographical maps. The two-way dielectric absorption in the ice is calculated using a measured temperature profile in the borehole. It indicates the reflection coefficient at the ice-bedrock interface to be very low. A list of satellite-obtained locations of map control points is included.

Introduction

Radio-echo soundings have been made in the Dye 3 area in 1971, 1978, and 1979 with airborne equipment [Gudmandsen and Christensen, 1972; Overgaard, 1978; 1979]. In 1975 soundings were made from the surface [Sondergaard, 1975] along the strain net established by S. Mock (Mock, personal communication). These soundings gave an overall picture of the bedrock and surface topography in the area [Overgaard and Gudmandsen, 1978].

These surveys have shown that the bedrock topography in the Dye 3 area is very mountainous and much like the rest of the southern part of Greenland. Due to the complicated flow over the hilly bedrock at Dye 3, it is difficult to calculate a time scale for the deep core. In order to perform these calculations with the best possible bottom profile, detailed soundings were made in a network that includes the flowline from the drill site upstream to the ice divide.

Essential requirements for reconstructing the subglacial landscape are:

Measurements of the ice thickness;

Measurements of the surface topography elevation;

Navigation—i.e. accurate locations for the above.

Since it is known that the bedrock topography fluctuates greatly within distances of less than 1 to 2 km in this subsurface mountainous part of Greenland, navigation accuracy of some 100 m is needed. This accuracy is much higher than can be carried out in ordinary airborne work, where the navigation is normally accomplished within 1 to 5 km.

The Radio-Echo Sounders

The principle of a radio-echo sounder is described by Evans and Smith [1969]. In the current work we have used two sounders, one for the surface work (TUD mk.I) [Christensen et al., 1970] and a more powerful one (TUD mk.II) [Skou and Sondergaard, 1976] for airborne operation.

Specifications of the two systems are given in Table 1. In the TUD mk.II sounder, distances measured on the recorded profile depend on the ground speed of the aircraft. In the TUD mk.I surface sounder, the pulse repetition frequency is controlled by a bicycle wheel odometer so that distances measured on the recorded profile have a linear relationship with the distance run by the sledge. Both sounder display systems work in the Z-mode with the ice thickness appearing as a profile on 35 mm film. In the airborne case, a marker is placed on the film at 15 second intervals and the aircraft speed and position are recorded. In the surface case, the film is fixed and the oscilloscope beam is moved stepwise with impulses from the odometer. Each image covers approximately 4 km.

The Z-mode signals are differentiated, therefore the forefront of the signals is pronounced. An example of a Z-profile is shown in Figure 1. The sweep is started 2 μsec before the radar pulse is transmitted in order to reduce possible nonlinearities in the display system. In the upper part of the ice, several layers are visible. The 2 μsec pulse train to the right is used to calibrate the depth reading.

TABLE 1. Main Parameters for the Radio-Echo Sounders mk. I and II, Where the Sounder mk. I is Used for Surface Work and the Sounder mk. II is Used for Airborne Work.

	TUD mk. I	TUD mk. II
operating frequency	60 MHz	60 MHz
peak power	1.3 kW	10 kW
pulse length	1 μsec.	0.25 μsec.
receiver bandwidth	1 MHz	4 MHz
system sensitivity	175 dB	205 dB
range accuracy	15 m	15 m
antenna	3 elem. Yagi	four dipoles

[1] Now at: Geophysical Isotope Laboratory, University of Copenhagen, Denmark.

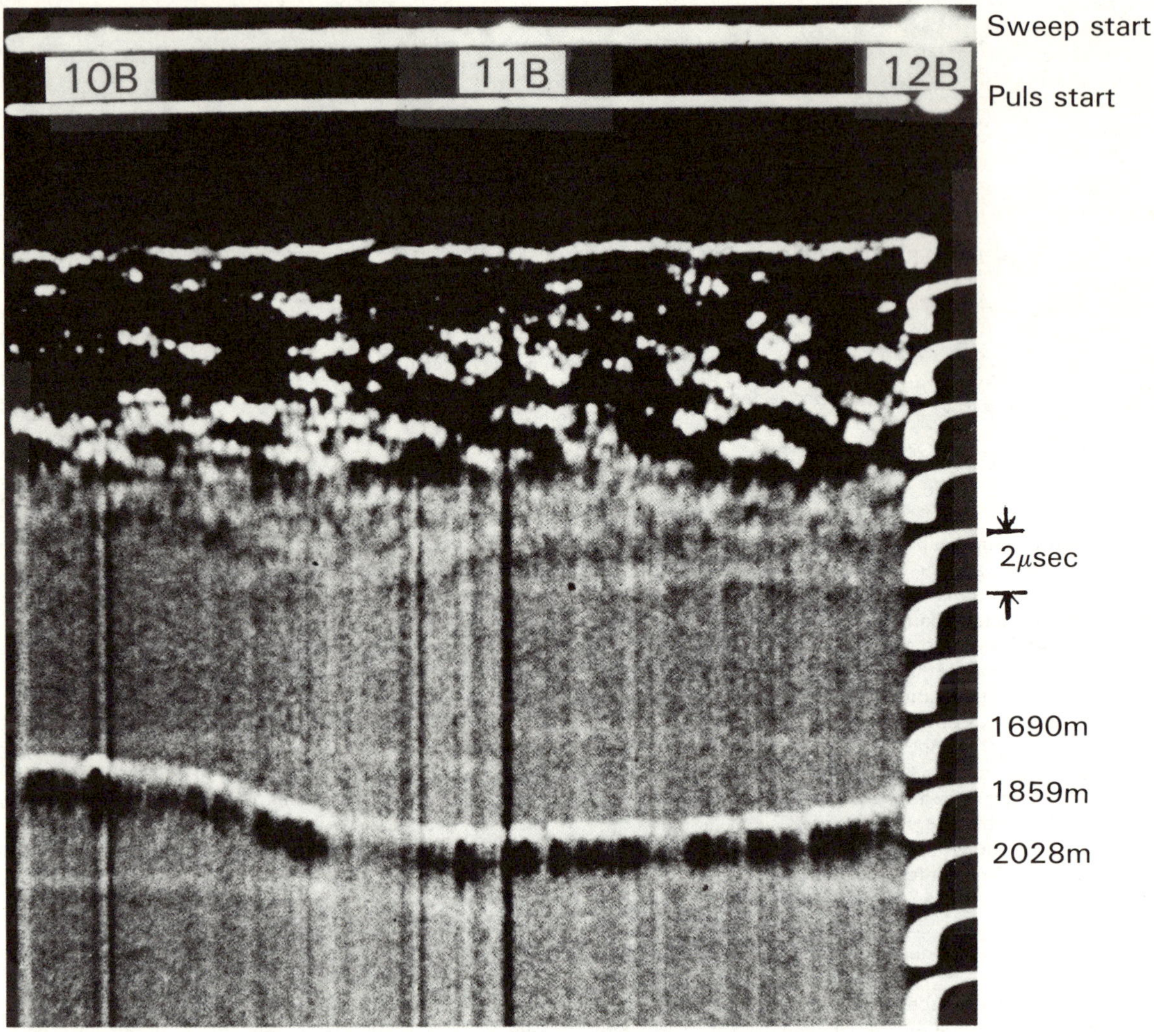

Fig. 1. Z-scope picture of a 4 km profile 16 km upstream from 1B.

Navigation

Surface radio-echo sounding

During the surface measurements, navigation was done by the U.S. Navy Navigation Satellite System using a geodetic receiver. This system measures the position of a receiver antenna in a geocentric coordinate system. Thus, one obtains the height of the antenna relative to the ellipsoid WGS-72 and the geographic coordinates. The three coordinates may be determined with an accuracy less than 2 m, providing the receiver is stationary for some days. In order to obtain the height above mean sea level (MSL), the geoid height must be subtracted from the height above the reference ellipsoid. In the present work, a geoid height of 40 m has been used [Rapp, 1978; Tscherning and Forsberg, 1981].

The soundings were carried out along lines with their end points fixed by the geoceiver. Each half kilometer was marked by a pole, and all intermediate marker points can be given geographic coordinates by the odometer readings. The main error in the navigation is the drift and the nonlinearity of the display system. This error is estimated to be less than 100 m.

The major part of the surface soundings followed the strain net established by the Ohio State University. The main purpose of this strain net was to determine the flowline from the deep bore hole and 20 km upstream. The first station of the strain net is located as close as possible to the deep hole (180 m). The azimuth of the network is 60.4° which is the same as the measured direction of the ice velocity 450 m from the deep hole (Table 2). We conclude that the center line of the strain net (the B-line) is a good approximation to the flowline. Upstream, the surface elevation increases southwards [Reeh et al., 1984], therefore the flow line must turn in this direction. Since the only determination of the flow line upstream is stations 3005 and 3007 in Figure 2 [Drew, 1982] the flow line indicated in Figure 2 is tentative [Reeh et al., 1985]. A list of the geoceiver stations established is given in Table 2.

Airborne radio-echo sounding

The aircraft was equipped with a pressure altimeter and an Inertial Navigation System (INS) that gave the position with an accuracy of 1–5 km. The errors in the INS system are accumulated during the flights. However, corrections could be made if more than one reference point was passed during the flight.

In the present work three passes over Dye 3 have been used, one

TABLE 2. Measured Geographic Coordinates and Heights Above The WGS-72 Ellipsoid (semimajor axis 6378135 m, flattening 1/298.26). (The coordinate are given in degrees and the height in meters above the WGS-72 reference ellipsoid. The height above mean sea level is assumed to be 40 m less than the indicated height.)

Site		Latitude (N)	Longitude (W)	Height
Drill site	+	65.18621	43.81975	2529
Site Iben	+	65.00921	44.69240	2673
Site Helens	+	65.25925	43.74267	2494
Site Sct. Hans	+	65.06814	43.94008	2565
12 C	+	65.08271	44.17612	2599
1 C	+	65.17218	43.80127	2530
16 B	+	65.06367	44.34052	2639
13 B	+	65.08971	44.23071	2600
9 B	+	65.12272	44.09115	2575
5 B	+	65.15631	43.94837	2560
1 B	+	65.18731	43.8227	2524
−7 B	+	65.21254	43.6637	2485
3007 (A-line)	*	65.01414	44.64322	2675
3005 (A-line)	*	65.11869	44.19755	2600
5 A	+	65.17208	43.96536	2546
1 A, 3003	*	65.20152	43.84021	2530

+ Measured by Geophysical Isotope Laboratory, University of Copenhagen.

* 1980 positions from Drew, [1982].

In addition, the position of Station 1 [Mock, 1976] located 450 m from the deep drill site was measured in July 1981 to 65.187878°N, 43.828382°W, height 2529.68 m. Combined with the previous measurements, this corresponds to a surface velocity of 12.32 ± 0.35 m/yr, direction 61.15 ± 1.8°.

pass in the beginning, a second one in the middle and a third at the end of the flight. These passes are clearly visible on the recorder film as echoes from the station (hyperbolic in time), picked up by the antenna. By knowing the flight altitude relative to the surface the minimum distance to the station can be calculated accurately. This procedure allows for calibration of the navigation data, the positional errors being less than 300 m and the height errors being less than 20 m. These values are confirmed by checking the data at all cross points. A map showing all measurement lines from airborne and surface measurements is given in Figure 2. Differences in accuracy and data density of the surface and air surveys are taken into account by suitable weighting [DK.IDIMS, 1982]. The result is a digital model with a horizontal resolution (pixel size) of 190*190 m.

Measurements of Ice Thickness

There are some corrections to be made to the thickness measured by the radar. Firstly, the velocity of the electromagnetic wave of the ice depends on the density of the ice. In solid ice with a density of 920 kg/m^3 the velocity is 169 m/μs [Evans, 1965]. Although ice velocities from 168 m/μs to 173 m/μs have been published [Drewry et al., 1982], 169 m/μs is assumed in all Danish soundings in Greenland. As the density in the firn is lower than in pure ice, the velocity will be higher, and a correction of +10 m has been calculated based on density profiles from the Dye 3 area [Herron and Langway, 1980], assuming a linear relation between permittivity and density.

The second correction allows for the length of the cable between the transmitter and the antenna, because the transmission time is measured from the transmitter. The signal velocity in a polyethylene insulated coaxial cable is 200 m/μsec which gives the "equivalent ice length" of a standard cable

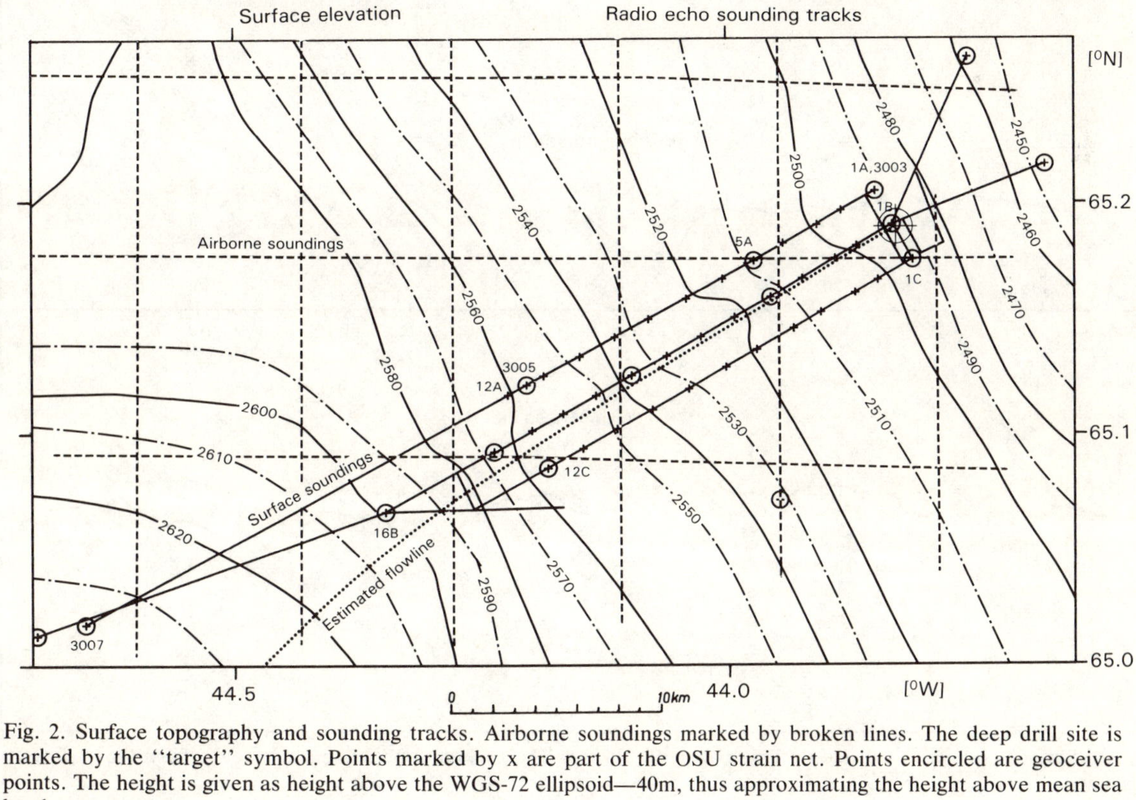

Fig. 2. Surface topography and sounding tracks. Airborne soundings marked by broken lines. The deep drill site is marked by the "target" symbol. Points marked by x are part of the OSU strain net. Points encircled are geoceiver points. The height is given as height above the WGS-72 ellipsoid—40m, thus approximating the height above mean sea level.

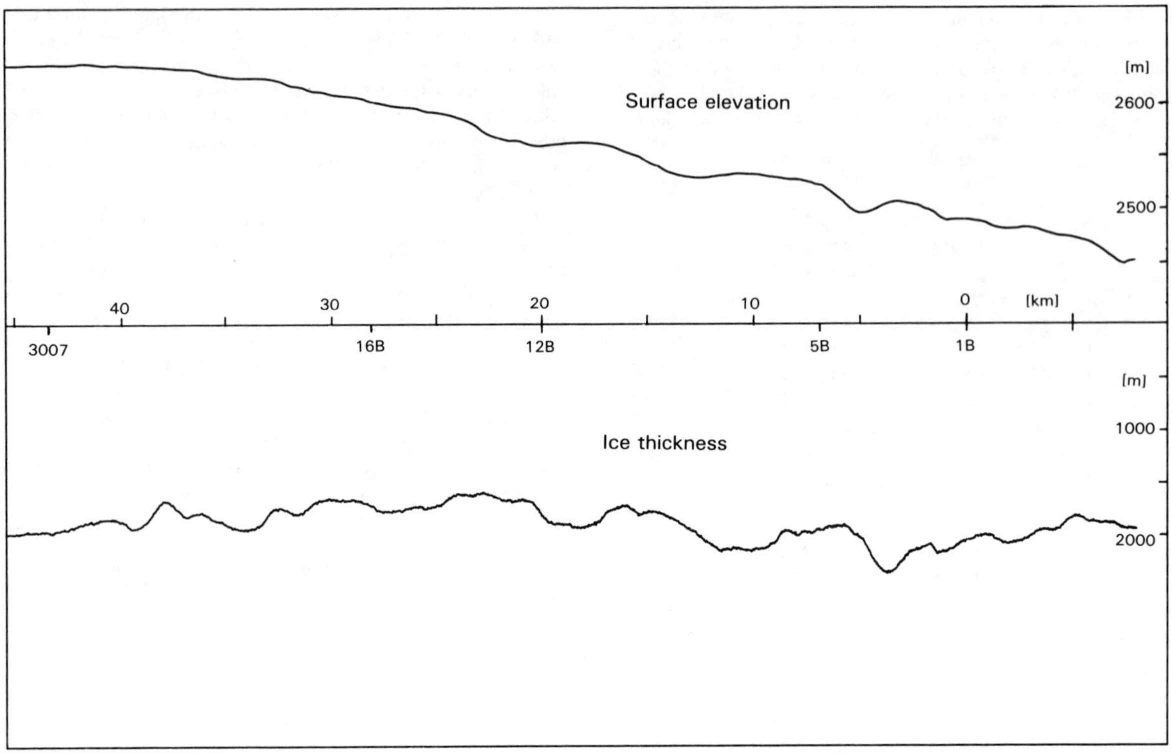

Fig. 3. Surface elevation and ice thickness along the B-line. Elevations are in meters above mean sea level (MSL).

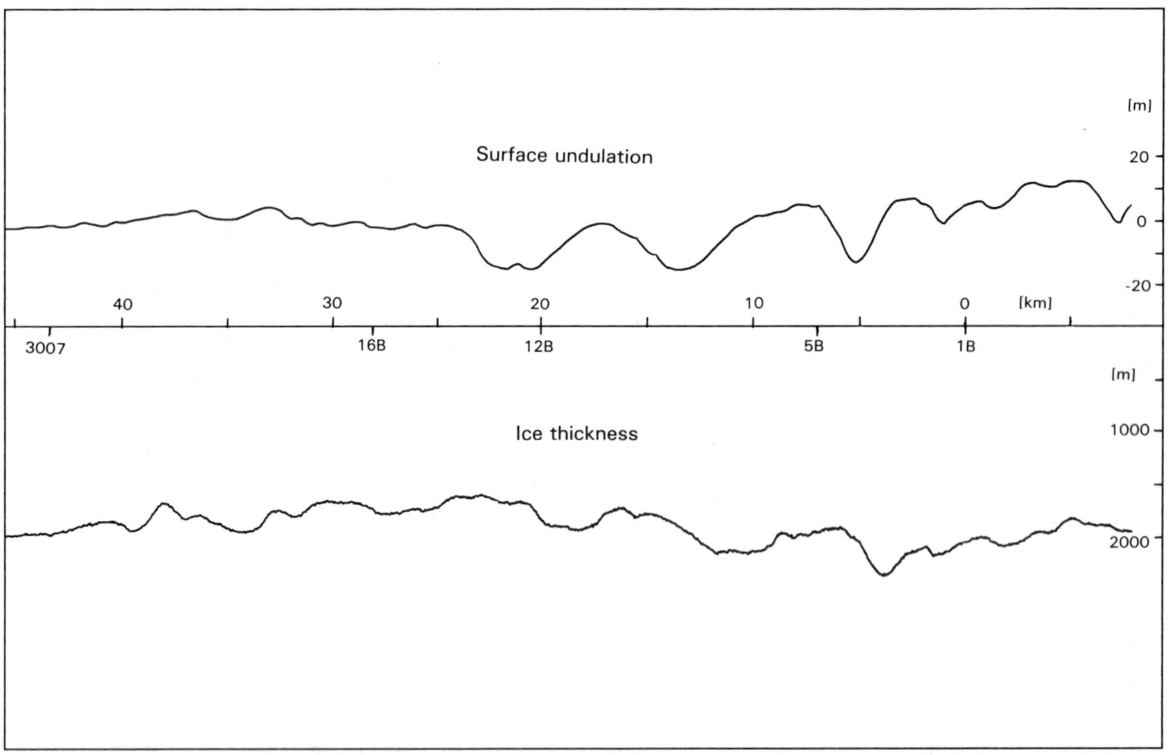

Fig. 4. Surface undulation and ice thickness along the B-line. The general trend has been removed from the surface elevation.

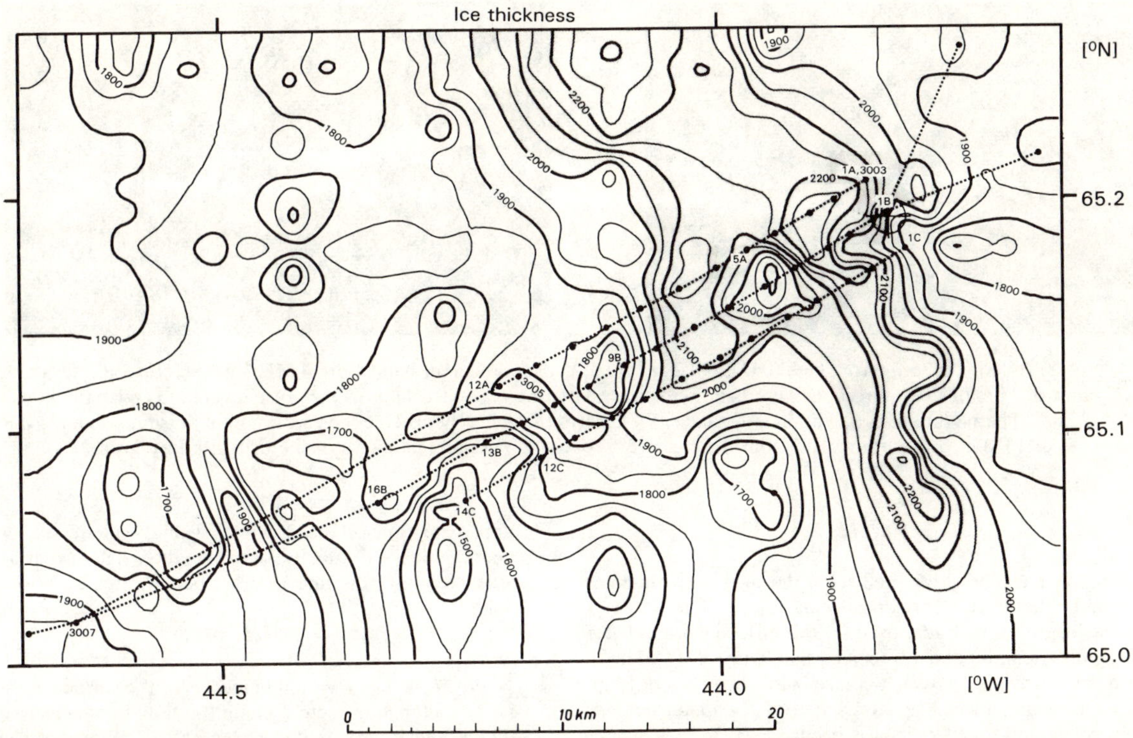

Fig. 5. Contour map showing the ice thickness. The contours are drawn with 50 m intervals. The drill site is marked by a "target" symbol. The most important surface tracks are indicated by broken lines. Ice thicknesses greater than 2000 m are shaded.

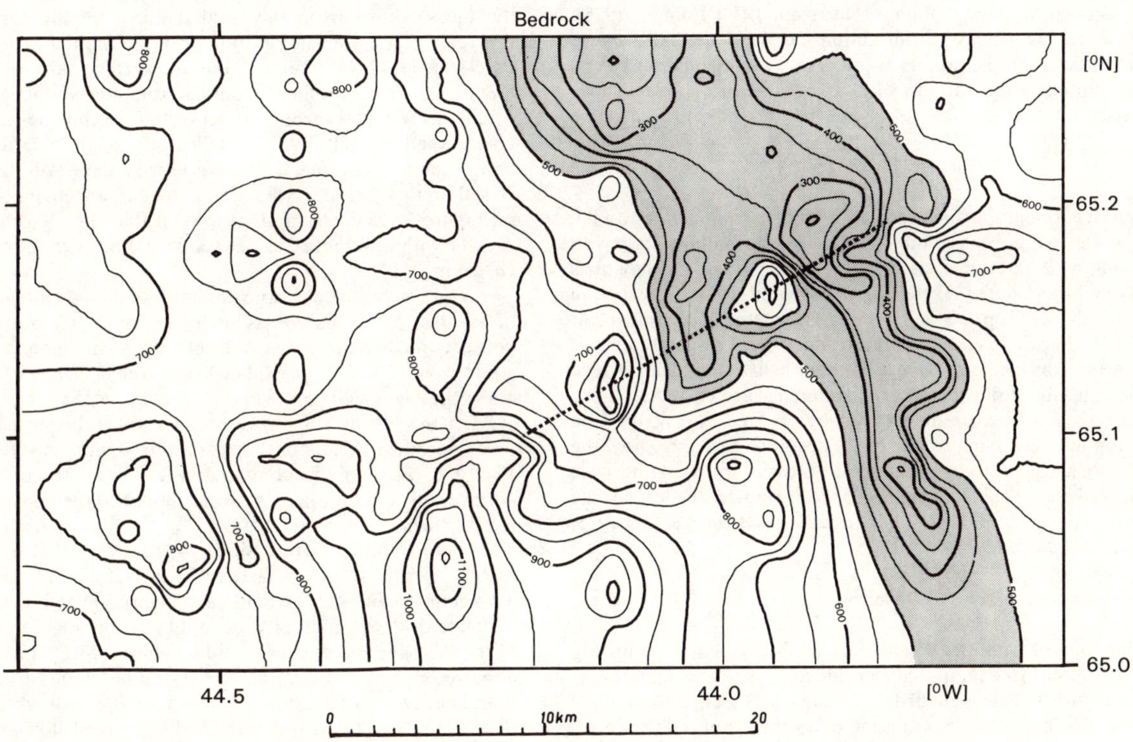

Fig. 6. Contour map showing the bedrock height in meters above sea level. The contours are drawn with 50 m intervals. The B-line from 1B to 12B is marked by a broken line. Bedrock heights below 500 m are shaded.

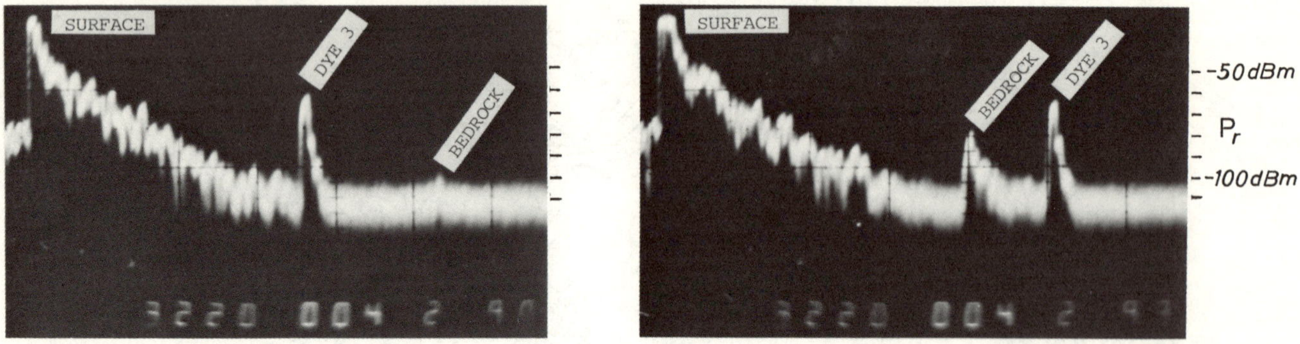

Fig. 7. A-scope records from flight 0478, pulse length 0.25 μsec, receiver bandwidth 4 MHz. The picture at left, at a distance of 3 km upstream from the Dye 3 station, shows a bedrock echo 5 dB above the noise level, and an ice thickness of 2225 m. The picture to the right, at a distance of 4 km from Dye 3, a bedrock echo 25 dB above noise, and an ice thickness of 1650 m. Both frames are from the east-west sounding passing just south of the drill site in Figure 2.

$$L_{ice}=0.845L_{cable} \quad (1)$$

This correction must be subtracted from the measured thickness. With a 12 m cable the two corrections cancel each other.

The pulse length for the radar used for the surface work is 1 μsec. With this pulse length, and a receiver bandwith of 1 MHz, the receiver delay corresponds to an increase in ice thickness of 10 m to 30 m for a strong and a weak signal respectively [Christensen et al., 1970]. No correction for this effect is used.

Finally, the radar measures the first arrival of an echo. In complicated terrain, this echo does not necessarily originate from the point directly below the antenna. As a result, the profiles are smoothed, and narrow steep valleys may not be visible. This error may be reduced by deconvolution [Harrison, 1970; French, 1975]; however in our case no such correction has been applied because the distance between the profiles is too great. The high directivity of the Yagi antenna used for the surface work reduces side echoes [Robin et al., 1977].

Surface Topography

The surface topography is measured by pressure altimeters. In the surface work, two pressure altimeters are used simultaneously with one serving as a reference station at a place where the height is known from other sources (geoceiver fixed points). Thus correction can be made for all errors other than the spatial variation in air pressure. Within distances of 20 km, under stable weather conditions, errors in surface elevation were kept below 2 m. The surface elevations measured during airborne soundings are corrected using the surface-based measurements as a reference. The result is a surface topographical map (Figure 2) accurate to 3 m when the positions are close to the surface based measurements. Otherwise errors degrade to 20 m. The map is in agreement with the large-scale satellite-based topography of South Greenland [Brooks et al., 1978; Zwally et al., 1983].

Results

The data output from the sounder are profiles (Figure 1) showing the ice thickness. An example of a profile after the data treatment is shown in Figure 3. This is the B-line profile starting approximately 2 km west of the ice divide, passing 150 m northwest of, and ending 8 km northeast of the deep drill position. The line 12B to 1B is a good approximation of the flowline. Table 2 identifies the various site geographical coordinates and heights. The general trend of the surface along this line is described by the equation [Reeh et al., 1985]

$$H = H_o - ax^n \quad (2)$$

where H_o is the elevation in meters of the surface at the ice divide, x is the distance in meters along the flow line measured from the ice divide and $a = 1.2710^{-4}$ and $n = 1.32$. Figure 4 shows the same profile as Figure 3 with expanded height scale, but the general trend has been subtracted from the surface, so that only local features of the surface topography appear. From Figure 4 it is seen that the main features of the bedrock topography, as reflected in the ice thickness, show up in the local features on the surface with an upstream phase shift of 90° of the bottom undulations [Budd, 1970; Paterson, 1981] resulting in maximum surface slope at minimum ice thickness. From profiles obtained from the surface and airborne soundings a digital terrain model of the ice thickness in the area is constructed by splitting the profiles into single records and recombining them according to the navigation data (DK.IDIMS. 1982). Based on this digital terrain model the contour map of ice thickness on Figure 5 is constructed. It is seen that the ice thickness varies rapidly with differences in excess of 500 m over short distances of ca. 2 km.

In areas with just airborne soundings, such as the northwest part of the map, it is not possible to reconstruct a satisfactory map because the distance between tracks are 8 km compared to a typical structure of 2 km. This explains the circular shape of the contours along the flight tracks.

After the ice thickness is subtracted from the surface, a bedrock model ensues. A contour map of this digital model is shown in Figure 6. Due to the comparatively flat nature of the ice-sheet surface, this contour map is rather similar to the contour map of the ice thickness. The major features on this map are, firstly, the drill site is close to the eastern slope of the valley that trends approximately northwest - southeast. If the drill site had been moved only 3.5 km upstream, there would have been 2300 meters of ice through which to drill! Secondly, it is seen that further upstream to the north of the flow line there is a flat and level area with an ice thickness of about 1800 m, and in contrast, a very mountainous area south of the flow line. Due to the complex nature of the bedrock topography it may be necessary to use a three-dimensional flow model to construct a time scale for the deepest part of the core.

In the radar recordings a weak bedrock echo is obtained from the valley area west of the drill site. The dielectric absorption of the ice can be calculated if the temperature profile of the ice column is known [Gudmandsen, 1971]. This was measured in the drill hole in 1982 and the two-way absorption was determined as 69.2 dB. This figure can then be used to calculate the received power by using the radar equation [Gudmandsen, 1971]

$$P_r = (1/64\pi^2) \times G_t^2 l^2 L_{rr} L_{ti}^2 P_t \times q_f (h + d)^2 L_i, \qquad (3)$$

where P_r is the received power, G_t the antenna gain, l the wavelength, L_{rr} the power reflection coefficient of the ice-bedrock interface, L_{ti} the power transmission factor at the air-ice interface, L_i the dielectric absorption in the ice, P_t the transmitted power, h the distance between the airplane and the ice surface, d the thickness of the ice, and q is the refraction gain. The received power should be compared to the receiver noise. Neglecting the antenna noise, the receiver noise is given by

$$P_n = FkT_oB, \qquad (4)$$

where F is the receiver noise figure, k is Boltzmann's constant, T_o is the absolute temperature of the receiver and B is the bandwidth of the receiver. Using the parameters for the airborne radar, TUD mk. II, the dielectric absorption in the ice, a value for L_{rr} of -11 dB ($e_r =$ 10), and a flying height of 600 m above the ice, we can find in the valley 3 km upstream of the drill site a received power of 30 dB above the noise level of the receiver. The corresponding A-frame is the left picture in Figure 7. The bedrock echo is 5 dB above the noise level or 25 dB less than calculated. Four kilometers southeast of the drill site, the calculated echo strength is 51 dB above the receiver noise level. The corresponding A-frame is to the right in Figure 7, indicating a signal 25 dB above noise, or 26 dB less than calculated. Based on the Z-film, these values are representative of the bottom echo in the regions. A possible explanation for this general 25 dB too weak echo could be:

In the calculation of the received power the assumption is that the bedrock is a plane surface and that the real part of the permittivity, e_r, of the bedrock has a value of 10. However the bedrock is not planar but will have a certain degree of roughness causing a signal reduction due to scattering. An estimate of this loss is 12 dB [Gudmandsen, 1971]. Furthermore, at the drill site, the last 25 m of the core was found to be silty ice [Gundestrup et al., 1985]. This means that the permittivity of the medium may change gradually from that of ice to that of the bedrock, and not abruptly, as assumed for the above calculations. Compared with a radar wavelength in ice of 2.8 m, this will lower the power reflection coefficient for the ice-rock interface, L_{rr}, significantly. A thick layer of moraine such as found at Camp Century in north Greenland [Hansen and Langway, 1966] may further reduce the echo. Also, the dielectrical loss in the ice may be higher than assumed. Combined, these effects may account for the difference of 25 dB. Further work is needed in order to separate the components.

Acknowledgments. The major part of the surface measurements followed the strain net established by I.M. Whillans of Ohio State University. Satellite precise ephemeris has been provided by Defense Mapping Agency Topographic Center.

This work has been sponsored by the Commission for Scientific Research in Greenland, Copenhagen; the Danish Natural Science Research Council; and the U.S. National Science Foundation Division of Polar Programs, Washington, D.C.

References

Brooks, R. L., W. J. Campbell, R. O. Ramseier, H. R. Stanley, and H. J. Zwally, Ice sheet topography by satellite altimetry, *Nature, 274,* 539–543, 1978.

Budd, W. F., Ice flow over bedrock perturbations, *Journal of Glaciology, 9,* 29–48, 1970.

Christensen, E. L., N. Gundestrup, E. Nilsson, and P. E. Gudmandsen, Radioglaciology: 60 MHz Radar, *R77,* Electromagnetics Institute, Technical University of Denmark, Lyngby, 1970.

DK.IDIMS, *TMPAK Users Guide,* Electromagnetics Institute, Technical University of Denmark, Lyngby, 1982.

Drew, A. J. R., Glacial movements in Greenland from doppler satellite observations, M.S. thesis, the Ohio State University, Columbus, 1982.

Drewry, D. J., S. R. Jordan, and E. Jankowski, Measured properties of the Antarctic ice sheet: Surface configuration, ice thickness, volume and bedrock characteristics, *Annals of Glaciology, 3,* 83–91, 1982.

Evans, S., Dielectric properties of ice and snow—a review, *Journal of Glaciology, 5,* 773–792, 1965.

Evans, S., and B. M. E. Smith, A radio echo equipment for depth sounding in polar ice sheets, *Journal of Scientific Instrumentation, Series 2, 2,* 131–136, 1969.

French, W. S., Computer migration of oblique seismic reflection profiles, *Geophysics, 40,* No. 6, 961–980, 1975.

Gudmandsen, P. E., Electromagnetic Probing of Ice, in *Electromagnetic Probing in Geophysics,* edited by J. R. Wait, pp. 321–348, Galem Press, Boulder, 1971.

Gudmandsen, P. E., and E. L. Christensen, Radioglaciology: Report on Activities in the Period September 19–October 3, 1971, *D141,* Electromagnetics Institute, Technical University of Denmark, Lyngby, 1972.

Gundestrup, N. S., S. J. Johnsen, and N. Reeh, ISTUK a deep ice core drill system, Second International Symposium on Ice Drilling Technology, August 30–September 1, 1982, Calgary, Canada, *Special Report,* U.S. Army Corps of Engineers, Cold Regions Research and Engineering Laboratory, Hanover, N.H., in press, 1985.

Hansen, B. L., and C. C. Langway, Jr., Deep core drilling in ice and core analyses at Camp Century, Greenland, 1961–1966. *Antarctic Journal of the United States, September–October,* 207–208, 1966.

Harrison, C. H., Reconstruction of subglacial relief from radio echo sounding records, *Geophysics, 35*(6), 1099–1115, 1970.

Herron, M. M. and C. C. Langway, Jr., Firn Densification: An empirical model, *Journal of Glaciology, 25*(93), 373–385, 1980.

Mock, S. J., Geodetic positions of borehole sites of the Greenland Ice Sheet Program, *Report 76-41,* U.S. Army Corps of Engineers, Cold Regions Research and Engineering Laboratory, Hanover, N.H., 1976.

Overgaard, S., Radio Echo Sounding: GISP-1978 Field Report, *AR 56,* Electromagnetics Institute, Technical University of Denmark, Lyngby, 1978.

Overgaard, S., and P. E. Gudmandsen, Radioglaciology: Surface and Bedrock Contour Maps at Dye-3, *R 199,* Electromagnetics Institute, Technical University of Denmark, Lyngby, 1978.

Overgaard, S., Radio Echo Sounding: GISP-1979 Field Report, *AR 72,* Electromagnetics Institute, Technical University of Denmark, Lyngby, 1979.

Paterson, W. S. B., *The Physics of Glaciers,* Second Edition, Pergamon Press, Ltd., New York, 164–167, 1981.

Rapp, R. H., A. Global 1 deg. × 1 deg. Anomaly Field Combining

Satellite, Geos-3 Altimeter and Terrestrial Data., Department of Geodetic Science, *Report No. 278,* The Ohio State University, Columbus, Ohio, 1978.

Reeh, N., S. J. Johnsen, and D. Dahl-Jensen, Dating the Dye-3 deep ice core by flow model calculations, this volume, 1985.

Robin, G. de Q., D. J. Drewry, and D.T. Meldrum, International studies of ice sheet and bedrock, *Philosophical Transactions of the Royal Society of London, B. 279,* 185–196, 1977.

Skou, N., and F. Sondergaard, Radioglaciology: A 60 MHz Ice Sounder System, *R 169,* Electromagnetics Institute Technical University of Denmark, Lyngby, 1976.

Sondergaard, F., Radioglaciology: Surface Soundings near Dye-3, *D 258,* Electromagnetics Institute, Technical University of Denmark, Lyngby, 1975.

Tscherning, C. C., and R. Forsberg, Geoid-determinations in the Norwegian Greenland Sea: An Assessment of Recent Results., Presented at IV Discussion Meeting on the Blue Road Geotraverse, Berlin Free University, February 11–13, 1981.

Zwally, H. J., R. A. Bindschandler, A. C. Brenner, T. V. Martin, and R.H. Thomas, Surface Elevation Contours of Greenland and Antarctic Ice Sheets, Journal of Geophysical Research, *88*(C3), 1589–1596, 1983.

DATING THE DYE 3 DEEP ICE CORE BY FLOW MODEL CALCULATIONS

N. Reeh, S. J. Johnsen[1], and D. Dahl-Jensen

Geophysical Isotope Laboratory, University of Copenhagen, Copenhagen N, Denmark.

Abstract. The primary objective of the ice-flow model calculations presented is to establish a steady state reference, to which measured data from the Dye 3 deep ice core can be compared in order to evaluate whether "irregularities" of the data are due to temporal variations. The influence of the rough topography of the ice sheet base upstream of the drill site and of the irregular distribution of the accumulation rate at the ice sheet surface are included in the flow model, which is based on perturbation techniques. Ice flow is divided into (1) a "basic" flow, accounting for the influence of the trends of the ice thickness and accumulation rate upstream from the drill site, and (2) a "perturbation" flow, accounting for the influence of the upstream perturbations of these quantities. Input data are: (a) the distributions along the flow line of surface and base elevations; (b) the distributions along the flow line of the accumulation rate; and (c) the temperature-depth profile. The dependence of the ice-flow properties on the stress and temperature fields (Glen's law) is taken into account. Flow-law parameters are derived from the modelling of the "basic" flow (giving values in reasonable agreement with hitherto derived values) and are used as input to the "perturbation" flow model. The model is used to calculate theoretical age and annual layer thickness profiles, which are compared with observed profiles along the Dye 3 deep core. The model seems to explain some, though not all "irregularities" of these profiles. Discrepancies are discussed in terms of imperfections of the model, e.g. that a simplified flowline course has been applied, that the model is a first order perturbation model and, that the ice flow may not be two dimensional and certainly not stationary.

Introduction

The primary objective of the ice-flow model calculations presented is to establish an accurate steady-state reference, e.g. for the time scale (the age-depth profile) or correspondingly the annual layer thickness profile (the λ-depth profile) along the Dye 3 deep ice core. This is important since irregularities of measured quantities along the core should not be interpreted as being due to temporal variations if they might be explained by careful steady state ice flow modeling. In particular this study includes an investigation of the effect of an irregular topography on the ice sheet base and an irregular distribution of the accumulation rate along the ice flow line. At the Dye 3 location such effects are expected to be significant, basal elevation changes exceeding 400 m within distances of 2 km, [Overgaard and Gundestrup, 1984] causing undulations of the ice sheet surface with wave heights of more than 20 m, which in turn give rise to short distance variations of the accumulation rate distribution.

Of course the ice-flow model calculations will also provide a theoretical time scale along the ice core. However the establishment of such a time scale is not considered a primary objective, since experience shows that absolute dating of the Dye 3 deep ice core is possible at least 6000 years—quite likely 8000 years, and possibly even 10,000 years—back in time; this dating will be more accurate than any dating provided by an ice flow model. Theoretical dating of ice older than 10,000 years by steady state flow modeling is of limited value since significant changes of the ice flow pattern are expected to have occurred during and in particular at the end of the glaciation, thus invalidating the assumption of steady state.

General Description of the Ice-Flow Model

The time scale and the λ-profile at a given location of an ice sheet are determined by the history of the upstream velocity field. For two-dimensional, steady-state ice flow, this velocity field in turn depends on (1) the distribution along the flow line of surface and base elevations, (2) the distribution along the flow line of the accumulation rate, (3) the distributions of ice temperature, ice crystal size, and ice fabric, and (4) the ice flow law parameters.

Two different approaches can be applied in deriving theoretical age and λ profiles:

(1) Analytical methods, applying a highly simplified model: Steady state, two dimensional flow, uniform upstream distributions of ice thickness and accumulation rate, no change of the temperature-depth profile in the direction of ice flow, and neglect of the influence of changing ice crystal size and ice fabric. [Haefeli, 1963; Dansgaard and Johnsen, 1969; Philbert and Federer, 1971]. Such models, which account for the influence of the general shapes of the velocity depth profiles, will be termed "sandwich" models, as a reflection of each annual layer having uniform thickness.

(2) Numerical methods, which furthermore take into consideration upstream variations in accumulation rate, base elevations and ice thickness, in some cases also horizontal ice flow divergence/convergence and even non-stationarity, see, e.g., Budd, [1969]; Johnsen, [1977]; Whillans, [1979]; Raynaud et al., [1979].

In spite of its simplicity, the analytical approach provides a good approximation to the smoothed time scale along the upper part of an ice core recovered not too far from an ice sheet summit or crest. On the other hand the numerical approach provides the best possible theoretical time scale at a particular location. This improvement, though, is achieved at the expense of increased model complexity and consequent costly computer calculations.

The ice flow model presented in this paper, belongs with the

[1] Now at: Science Institute, University of Iceland, Reykjavik, Iceland.

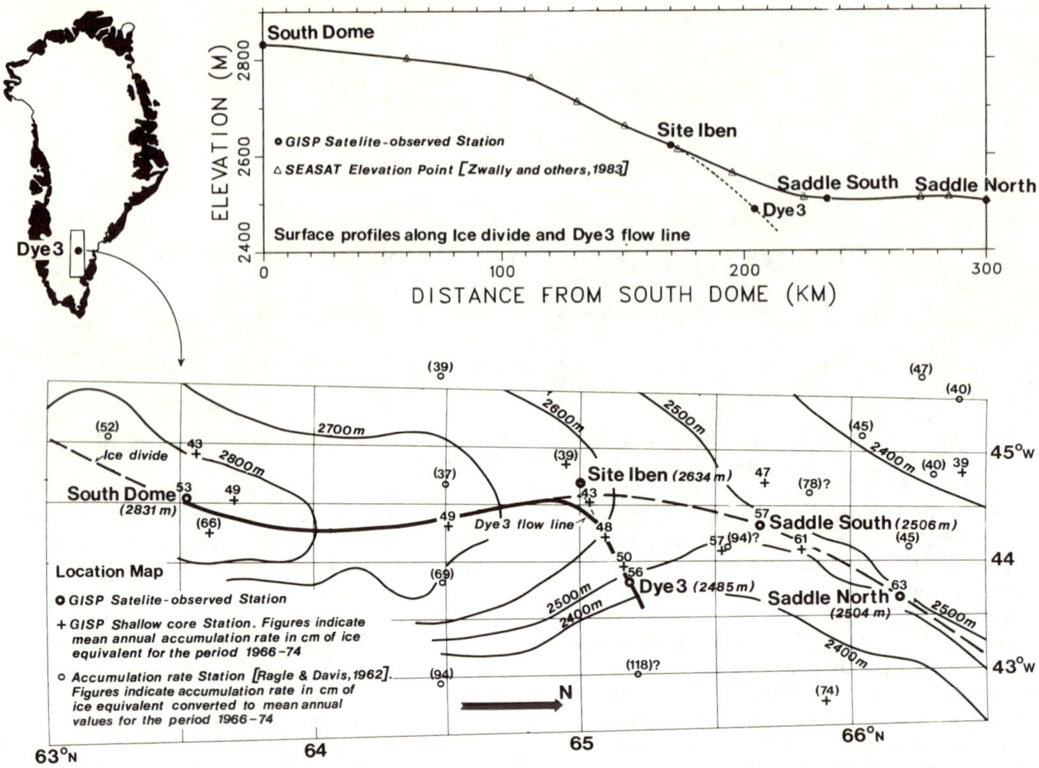

Fig. 1. Location map and surface elevation profiles of Dye 3 region. Elevations of GISP satellite stations are elevations relative to the WGS-72 ellipsoid reduced by 40 m to yield approximate elevations above mean sea level. SEASAT elevations are offset by a constant height decrement to match elevations of GISP stations.

numerical methods, the aim being to derive accurate age- and λ-profiles. Compared with previous numerical models however, improvements have been introduced at several points:

(1) The influence of the very irregular ice sheet base is accounted for by applying perturbation techniques, dividing the ice flow into a "basic" flow and a "perturbation" flow respectively. The "basic" flow model accounts for the ice flow over a smoothed base, with an ice flux determined by a smoothed accumulation rate distribution. The "perturbation" flow model accounts for the influence of the upstream deviations from the trend lines of these quantities.

(2) The derivation of the "basic" flow pattern is based on an approximate integration of the full system of flow equations, with due consideration of all stress components (shear stresses as well as longitudinal ones). Gradient terms in the direction of flow of the stresses and the vertical velocity component are however neglected, these terms being generally insignificant as compared with the leading terms of the equations. The influence on the ice-flow properties of a nonuniform temperature depth profile has been considered. Also the dependence of the ice flow properties on both the longitudinal stresses and the shear stresses through the effective shear stress (see Derivation of Ice Flow Law Parameters) is taken into account.

Thus the "basic" flow model provides a complete solution for the velocity, strain rate, and stress fields. The accuracy of the solution is checked by inserting the calculated fields in the full set of equations governing the ice flow, i.e. a) the equation of continuity, b) the equations of stress equilibrium and c) the ice flow law. Except close to the ice divide (i.e. within a distance of 1–2 ice thicknesses), the residuals are negligible, indicating consistency between the assumptions and the calculated "basic" solution.

One important quality of the "basic" solution is that it accounts for the change in shape of the horizontal and vertical velocity profiles, caused by the increasing influence on the ice-flow properties of the longitudinal stress deviator, as the ice divide is approached and the shear stress is accordingly reduced.

However, the influence on the ice flow properties of changing ice-crystal size and fabric with depth [Shoji and Langway, 1985; Herron et al., 1985] has not been considered in a systematic way, but rather by introducing a layer of low shear resistance at the base of the ice sheet. Furthermore, a very likely change of the ice fabric to a pattern more resistant to shear deformation as the ice divide is approached [Whillans, 1979] is not taken into account.

(3) The "perturbation" flow model is based on the theory originally presented by Budd [1970] for the velocity and stress fields associated with ice flow over an undulating base. Recently Hutter et al., [1981] gave a weighty contribution to the theory, by systematic application of perturbation theory. Using the ratio of the amplitude of the basal undulation to the ice thickness as the perturbation parameter, a hierachy of boundary value problems of increasing order (zero, first, second, etc.) was set up. Furthermore Hutter showed how the non-linear stress- and temperature-dependent flow law of ice can be considered within the framework of a systematic perturbation theory.

The perturbation flow model applied in this study is a first-order model based on the theory of Hutter et al. [1981]. At two points however, the approach has been modified. First, in Hutter's "basic" flow (zero-order flow) the vertical velocity and the longitudinal stress deviators vanish and the horizontal velocity does not vary in the direction of flow. No matter how necessary these assumptions are for a mathematically strict perturbation theory, they are equally poor approximations to the conditions in real ice sheets. Since the zero- and first-order solutions interact through the equations defining the first-order perturbation problem, it is more important to apply a realistic zeroth-order solution, rather than adhere to mathematical strictness. Therefore the local values of the zeroth-order velocity and stress solutions have been applied in the formulation of the first-order perturbation problem. Second, Hutter's correct treatment of the non-linear stress- and temperature-dependent ice rheology leads to a numerical algorithm that is rather time consuming, and consequently expensive to use in a computer program for flow-model calculations. For this reason a considerably faster approximate procedure has been developed: the idea is to approximate the depth variation of the stress- and temperature-dependent ice-flow law terms by exponential variations, thereby allowing an analytical solution to be obtained for the first order perturbation problem.

In this way the perturbation velocity field corresponding to a basal sine-shaped undulation of given wavelength and amplitude can be determined. In order to calculate the perturbation velocity field corresponding to the actual irregularities of the base and accumulation rate distribution, the basal irregularities are resolved in their Fourier components. For each component the corresponding perturbation velocity solution has been determined and these solutions have been added to the "basic" velocity solution to provide the compound velocity field.

The contribution to the perturbation velocity field from the irregular accumulation rate distribution is calculated on the assumptions that the accumulation rate variations are related to the surface undulations, display a constant phase shift in respect of these, and have an amplitude proportional to h/L, where h and L are amplitude and wavelength respectively of the surface undulation in question.

Next particle paths (which in the case of steady state flow coincide with the flow lines) and particle ages can be determined by numerical integration of the velocity field, yielding the ice-flow pattern and the time scale. Finally the λ profile is determined as the reciprocal of the derivative with respect to depth of the time scale.

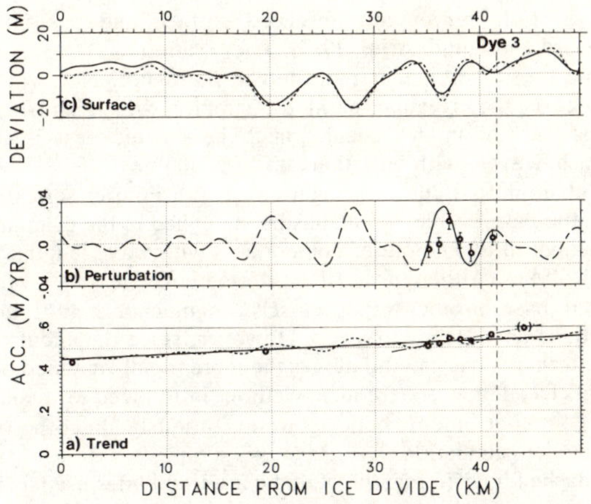

Fig. 3. Distribution of accumulation rate along the Dye 3 flow line. Points marked by circles are observed values. The various curves are explained in the text.

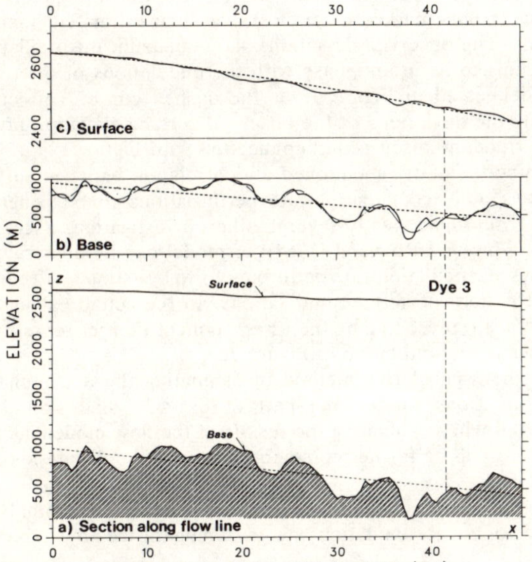

Fig. 2. Surface and base elevation profiles along the Dye 3 flow line. Full curves are observed, chain-dotted curves are smoothed, and dashed curves are trends of profiles.

The Data

The data used as input to the flow model are the following:

(1) Surface and base elevation profiles along the flowline, the course of which is supposed to be known.

(2) The distribution of accumulation rate along the flowline.

(3) The temperature depth profile.

The course of the flow line applied in the flow model calculations appears from the map in Figure 1, showing elevation contours and accumulation rate values of a sector of the south Greenland ice sheet. The thick curve is the estimated flow line through Dye 3, originating at the southern dome of the ice sheet some 200 km south of Dye 3. The flow line used in the flow-model calculations follows this curve upslope from Dye 3, but deviates from the thick curve as the ice divide is approached, continuing as a straight line to the ice divide close to Site Iben. One argument for choosing this simpler flow line course follows from comparing the mean surface slope along the ice divide at Site Iben with that along the flow line at Dye 3 (see surface elevation profiles shown in Figure 1). The ratio of these slopes is 0.46. Disregarding differences in ice thickness, ice temperature and ice fabric, the ratio of horizontal velocities goes with the slope raised to a power of 3. This means that velocities along the ice divide south of Site Iben are of the order of magnitude 0.46^3, ~ 10% of the velocity along the flow line at Dye 3 where the surface velocity has been measured to 12.3 m/yr [Overgaard and Gundestrup, 1984]. This deduction is confirmed by measured northgoing surface velocities at the ice divide in the order of magnitude of 2 m/yr [Drew, 1982]. So the flow along the south-north running ice divide is much slower than the flow along the east-west running part of the flow line from the ice divide to Dye 3. This means that the major part of the ice in the Dye 3 deep core (i.e. the upper 1700 m covering the last 8000 years) is expected to originate from near the latter flow line section where, furthermore, the most detailed topographic data and accumulation rate data have been obtained. The consequences of choosing an approximate course of the flow line will be further discussed in the section on annual-layer thickness, profile, and time scale.

The ice sheet topography along the selected flow line is shown in

Figure 2. Full curves are observed surface and base profiles [Overgaard and Gundestrup, 1985]. The position of Dye 3, some 41 km from the ice divide, is indicated in the figure. Noteworthy in Figure 2a is the irregular basal topography, which considerably impedes the flow model calculations. The ice thickness is around 2000 m, however, with variations up to ± 300 m.

The chain-dotted curve in Figure 2b is the Fourier series representing the base profile, which has been applied in the calculations. It is composed of 12 Fourier components with wave lengths in the range 4–50 km. Although the chain-dotted curve in general fits the observed base profile rather closely, significant high-frequency undulations have been cut away. However, since the effect on the flow pattern of a basal undulation is the more rapidly damped out the higher its frequency, the frequency cutoff, introduced for reasons of saving computer time, is believed to not seriously affect the results of the model calculations.

The dashed lines in Fig. 2a, b, and c are trend lines used as upper and lower boundaries for the "basic" flow model calculations. The surface trend line is determined as a least-squares fit to the observed surface-elevation profile, using a representation of the form

$$s = s_I + a\,x^b,$$

where x = distance from the ice divide.

The choice of a representation of this form is based on the fact that such an expression evolves from a power series expansion, in distance from ice divide, of the ice sheet profile equation of Nye, [1959]. The theoretical value of the exponent b is $(n + 1)/n$ (n = ice flow law exponent, see next section). The least squares fit gives b = 1.32, indicating an n-value close to 3. The base trend line is

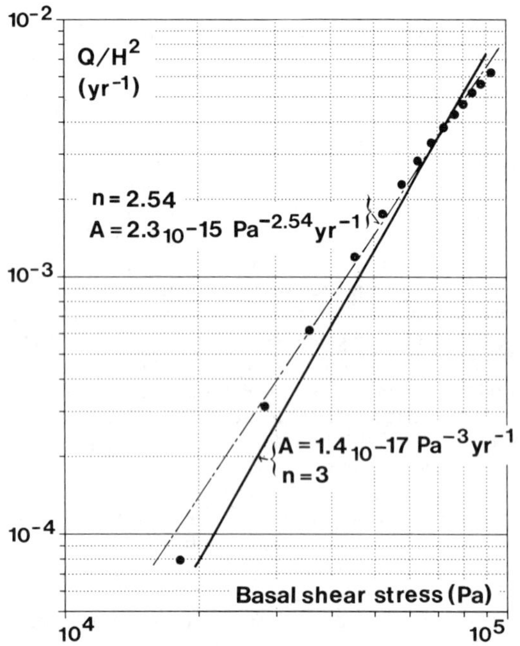

Fig. 4. Derivation of flow law parameters. Double logaritmic plot of ice flux Q divided by the square of the ice thickness H versus basal shear stress τ_b. Q, H and τ_b are "basic" values derived from the smoothed surface and base profiles shown in Figure 2b and c, and the smoothed accumulation rate distribution shown in Figure 3a.

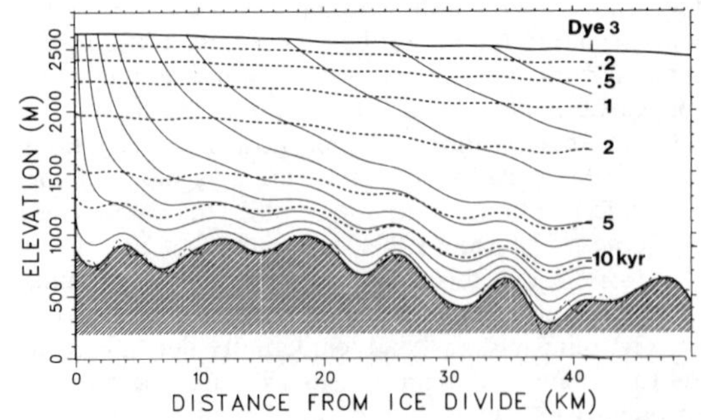

Fig. 5. Flow pattern predicted by the ice flow model. Thin full curves are particle paths (flow lines). Dotted curves are isochrones, with ages in thousands of years indicated to the right in the figure.

determined as the least squares linear fit to the observed base elevation profile.

The distribution of accumulation rate along the Dye 3 flow line is shown in Figure 3. In Figure 3a, points marked by circles are mean annual accumulation rates in m of ice equivalent for the period 1966–74, determined by stratigraphic analysis (i.e. by counting $\delta(^{18}O)$ seasonals) of shallow firn cores. The rightmost datapoint (at station 44 km) is less accurate than the other points: The accumulation rate record from this station covers the period 1962–70. However, a correction has been made to account for the difference of the mean annual accumulation rates for this period and the reference period 1966–74.

The full line in Figure 3a represents the general trend of the accumulation rate, which is used in the "basic" flow model calculations. A straight line has been chosen to represent this trend. The dashed curve in Figure 3a represents the estimated detailed accumulation rate distribution, which is more clearly illustrated in Figure 3b, showing the deviations of the detailed distribution from the trend line. Also shown in Figure 3b are the observed accumulation rate deviations (points marked by circles), and their estimated standard deviations. The observed deviations show a significant oscillation, which seems to be in antiphase with the undulations of the upper surface delineated in Figure 3c as the dashed curve. This curve represents the deviations of the observed surface elevation profile from the trendline discussed in connection with Figure 2.

Assuming the above mentioned anticorrelation between surface undulations and accumulation-rate perturbations to be generally valid, in accordance with several other investigations (see, for example, Gow and Rowland [1965]), a model can be constructed, that allows accumulation-rate perturbations to be estimated from the observed pattern of surface undulations. In the actual calculation this pattern is represented by the 12-component Fourier representation shown as the full curve in Figure 3c.

This rather speculative method of estimating the accumulation rate perturbations for a large part of the flow line should be remembered when evaluating the results of the flow model calculations. There will be further comment on this in the discussion and conclusions.

The ice flux variation along the flow line is determined simply by integration of the accumulation rate, assuming the ice flow to be two dimensional.

The temperature depth profile applied in the calculations is based on the 1980 temperature measurements in the Dye 3 deep hole down to 900 m depth. Down to this depth the temperature profile is approximately uniform (−20°C). Furthermore in 1981 the tempera-

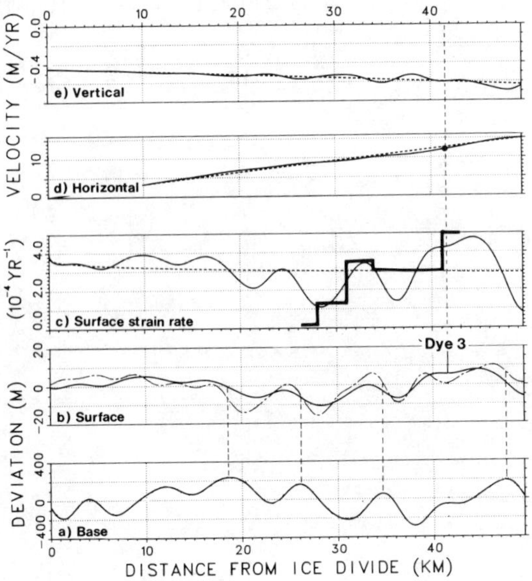

Fig. 6. Comparison of calculated surface undulations, strain rates, and velocities with observed values. a) 12-component Fourier representation of basal irregularities. b) Calculated (full curve) and observed (chain-dotted curve) surface perturbations. In c), d), and e) dotted curves and thin full curves are results of "basic" and "perturbation" flow model calculations respectively. The heavy full curve in c) is the surface strain rate distribution as measured by Mock, (unpublished data).

ture at the bottom of the drill hole was measured to −12°C by the thermistor built into the drill. Based on these data, a temperature profile for the entire depth of the ice sheet was estimated and used in the calculations. The deviation of this estimated profile from the complete temperature profile measured in the drill hole in June 1982 [N. Gundestrup, personal communication], is less than a few degrees Celcius. This difference will have a minor impact on the calculated velocity field. A recalculation with the application of the 1982-temperature profile will be performed later.

Derivation of Ice Flow Law Parameters

A detailed description of both the "basic" flow model and the "perturbation" flow model will be presented elsewhere. In this work only a few selected aspects will be considered in addition to the general description of the ice-flow model. One aspect to be mentioned is the derivation of the ice flow law parameters from the "basic" flow model, using the following expression, generalized from Budd [1969]

$$Q/H^2 = 2CA\tau_b{}^n, \qquad (1)$$

relating smoothed values of ice flux Q and ice thickness H to the basal shear stress $\tau_b = \rho g H \alpha$ (ρ = ice density, g = gravitational acceleration and α = surface slope). In equation (1) n and A are parameters of the ice-flow law (Glen's law, see, e.g., Paterson, [1981]),

$$\dot{\varepsilon} = A\tau_e{}^{n-1}\sigma',$$

where $\dot{\varepsilon}$ = strain rate, τ_e = effective shear stress and σ' = stress deviator.

The quantity C in equation (1) depends on the shape of the "basic" horizontal velocity profile and is accordingly, in principle, a function of the distance from the ice divide (see general description of the ice-flow model), as are also Q,H and τ_b. However, except in the immediate neighborhood of the ice divide, the variation of C is insignificant, and will consequently be neglected.

The flow-law parameter A is expressed as a fixed temperature, which, in the actual case, is the temperature at the upper surface of the ice sheet (−20°C).

Figure 4 shows a double logarithmic plot of Q/H^2 against τ_b. A least squares linear fit gives estimated values of $n = 2.54$ and $A_{2.54} = 2.3_{10}{-}15$ Pa$^{-2.54}$yr^{-1}, respectively. The values of n and A however are very sensitive to small changes of the ice flux Q, which unfortunately is encumbered with a rather large uncertainty, due to insufficient upstream accumulation rate and/or velocity data.

For this reason the flow-law exponent n is rounded to the value 3, and A is accordingly converted to units Pa^{-3}yr^{-1}, using a reference stress of 70,000 Pa. This gives $A_3 = 1.4_{10}{-}17$ Pa^{-3}yr^{-1}, corresponding to a temperature of −20°C. This A value is outside the range, applying to randomly oriented polycrystalline ice ($0.25_{10}{-}17$ − $0.57_{10}{-}17$ Pa^{-3}yr^{-1}) deduced from Paterson and Budd [1982], but well below the fabric-enhanced upper limit value of $2.8_{10}{-}17$ Pa^{-3}yr^{-1}, derived from the same paper. A comparison with the experimentally determined flow law parameters obtained by Shoji and Langway [1985], from measurements on ice core samples from the Dye 3 core, has not yet been performed.

Flow Pattern, Surface Perturbations, Surface Velocity and Surface Strainrates

As mentioned in the general description of the ice-flow model, the flow-law parameters derived from the "basic" flow model are used as input to the "perturbation" flow model. For each of the 12 Fourier components of the basal irregularities with wave lengths in the range 4–50 km, the corresponding perturbation velocity field has been determined and added to the "basic" solution.

Particle paths and particle ages can then be determined by numerical integration of the compound-velocity field, leading to the flow pattern and isochrone pattern shown in Figure 5. (Isochrones are surfaces of uniform age.) It appears that the isochrones reflect the shape of the bottom undulations, though with decreasing amplitude as the distance from the bottom increases.

Furthermore the undulations of the isochrones seem to be in phase with those of the bottom at least as high in the ice as they can be detected in Figure 5. Even though it cannot be recognized in Figure 5, due to the scale applied, a 90° phase shift occurs in the uppermost part of the ice sheet. This is illustrated in Figure 6 a and b, which show basal undulations as well as observed and predicted surface undulations. It appears from this figure that the theory predicts the shape and phase of the surface undulations reasonably well. The predicted surface amplitudes, however, are only about half as big as the observed ones. One might think of explaining this discrepancy as being due to using a two-dimensional (2d) "perturbation" flow model, although the actual basal topography is typically three dimensional (3D) [Overgaard and Gundestrup, 1984]. However, calculations applying a simplified three-dimensional perturbation flow model based on a linear ice rheology (to be presented elsewhere) indicate that 3D-surface amplitudes are decreased as compared to 2D ones. A possible explanation for the discrepancy is that an isotropic ice-flow law has been applied at all depths although the ice becomes more and more anisotropic as the bottom is approached, due to the development of ice fabrics with preferred vertical crystal axes orientation [Herron et al., 1985]. The resistance of this ice to stretching and compressing, which dominate the "perturbation" flow, is very likely higher than its resistance to shear deformation, which dominates the "basic" flow. Now, as previously mentioned, the "soft" ice flow parameters derived from the "basic" flow have been applied also for modeling the "perturbation" flow, which therefore very likely underestimates the amplitudes of the surface undulations.

A detailed presentation of the perturbation flow model, including a discussion of the velocity- and stress-fields, the transfer function and phase shift will be given elsewhere.

By means of the flow model it is also possible to theoretically predict the distribution of the surface strain rate and the surface velocities generated by the ice flow over the irregular base. In Figure 6c, d, and e such predictions are shown and compared with available observations. The observed strain rate distribution, shown by the thick curve in Figure 6c, is based on measured average surface-strain rates along a line with a slightly different course, than that of the flow line applied in this work (Mock, unpublished data).

The agreement between the observed and calculated strain-rate patterns seems to be satisfactory. However, more detailed strain-rate data obtained by Whillans (personal communication), confirm the theoretically predicted oscillating pattern of the surface-strain rate, but the observed amplitudes are twice the calculated ones. Since the amplitudes of surface undulations and surface-strain rates are linearly related [Whillans and Johnsen, 1983], this discrepancy is similar to the discrepancy of calculated and observed surface amplitudes, and may be similarly explained.

As regards the velocity distributions, one observation of the horizontal surface velocity provides a possibility for checking the model predictions. The data point marked by a circle in Figure 6d is obtained by repeated geo-ceiver observations during the period 1972–1982 of satellite station 1, established a few hundred m from the Dye 3 deep drill site by Mock [1976] but remeasured several times by N. Gundestrup. The measured velocity has a magnitude of 12.3 ± 0.4 m/yr [Overgaard and Gundestrup, 1985], which is in good agreement with the calculations. Also, surface velocities further upstream determined by strain-net measurements by Whillans (personal communication) compare well with the calculated velocities.

It should be noted that the basal undulations cause deviations of the horizontal surface velocity of up to 0.8 m/yr from the value predicted by the "basic" flow model. At the Dye 3 drill site the deviation has a magnitude of −0.6 m/yr. These deviations are of the order of 5–10% of the full values.

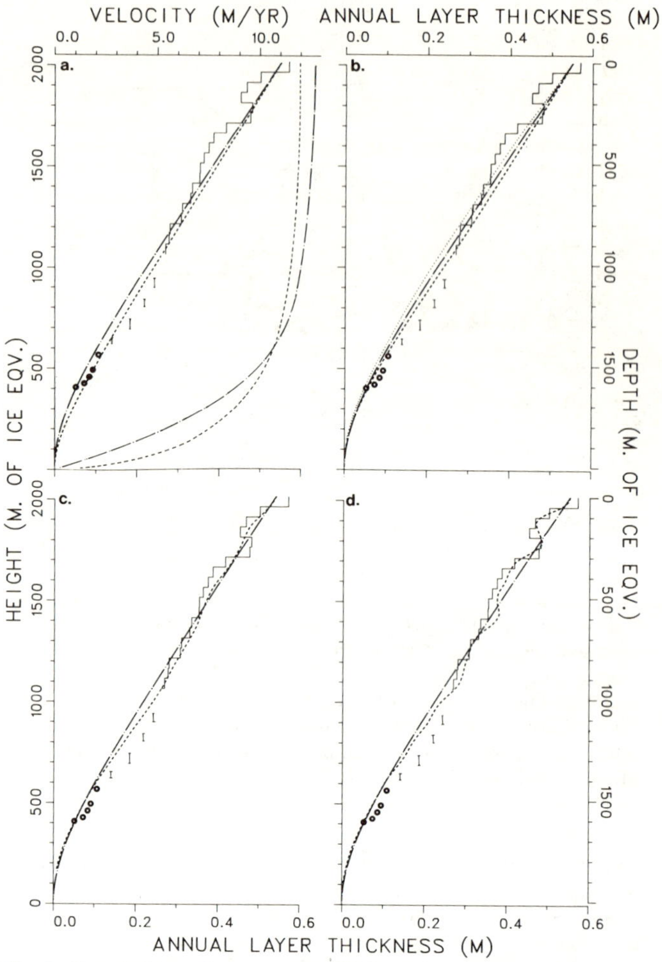

Fig. 7. Comparison of measured λ profiles with theoretical profiles obtained by models a, b, c, and d mentioned in the text. In all subplots the step curve, the line segments, and the points marked by circles are measured annual layer thicknesses. The chain-dotted λ curve, also shown in all subplots, applies to "Sandwich" model flow and is plotted as a reference profile. a) Shows two sets of horizontal-velocity profiles and λ profiles corresponding to "Sandwich" model flow, with and without a basal soft layer (the dashed and chain-dotted curves, respectively). b) The dotted curve is the λ profile for flow with uniform ice thickness but with a linear distribution of the accumulation rate corresponding to that of "basic" flow. The dashed curve is the λ profile for flow with the "basic" ice thickness distribution, but with uniform distribution of accumulation rate. c) Shows the λ profile for flow over the irregular base, but with a linear distribution of the accumulation rate corresponding to that of "basic" flow. d) Shows the λ curve for the full model, i.e. flow over an undulating base with irregular distribution of accumulation rate.

Annual-Layer Thickness, Profile, and Time Scale

The annual layer thickness profile (the λ profile) and the time scale (the age profile) are equivalent in the sense that they contain exactly the same information. They are just two different ways of representing the layer sequence along a vertical line in an ice sheet. Both are useful representations, the λ profile being the more suited for comparing measured and modeled sequences, and the age-profile being the most suited for interpreting ice core data.

Rather than just comparing the measured λ profile with the flow-model predicted profile, it is worthwhile performing a step by step comparison with profiles obtained from models of increasing complexity. Five models will be considered, (Figure 7):

a. Flow with uniform ice thickness and accumulation rate ("Sandwich"-model flow).

b_1. Flow with uniform ice thickness, but with a linear distribution of the accumulation rate, corresponding to that of "basic" flow, see Figure 3a.

b_2. Flow with a distribution of the ice thickness corresponding to that of "basic" flow (see Figure 2a) but with uniform distribution of accumulation rate.

c. Flow over an undulating base, but with a linear distribution of the accumulation rate, corresponding to that of "basic" flow.

d. The full model: Flow over an undulating base with consideration of the irregular accumulation-rate distribution.

By this step-by-step comparison, it becomes possible to study the separate effect on the λ profile of the various features.

To ensure a fair comparison, the accumulation rate and the ice thickness at the site, where the profiles are calculated (Dye 3), are adjusted to the same values for the various models, viz. 0.55 cm of ice/yr and 2037 m respectively. The latter value is equal to the length of the Dye 3 deep core, and corresponds to 2013 m of ice equivalent, when corrected for low density snow and firn in the uppermost 150 m of the ice sheet.

The step curve, the line segments, and the points marked by circles in each subplot of Figure 7, are measured annual layer thicknesses as obtained by measuring seasonals of $\delta(^{18}O)$, dust, and acidity, [Hammer et al., 1978; Hammer, 1980]. At the time of writing a continuous λ profile has been obtained down to 979 m depth (955 m of ice equivalent), containing 2600 annual layers and plotted in the subfigures as 50 m averages, each representing from 88 to 186 annual layers (the step curve).

The line segments shown in the subfigures represent λ values averaged over core sections, 20–50 m long, each containing from 160–280 annual layers. The points marked by circles are average λ's over shorter core sections, representing around 20 annual layers each.

In the case of "Sandwich"-model flow, the λ profile depends on the shape of the horizontal velocity profile only, since the accumulation rate and the ice thickness have been ascribed fixed values, as mentioned above.

In Figure 7a two theoretical λ curves are shown, corresponding to different velocity profiles having equal mean values, but one with, and the other without, consideration of a bottom layer of low shear resistance (the dashed and chain-dotted curves, respectively). The relative difference of the two λ profiles goes up to 30% around 1700 m depth. The corresponding difference of the time scales will be discussed in connection with Table 1. Neither of the theoretical profiles reproduce the irregularities of the measured profile, nor do they reproduce its trend. The evidence for ice fabric development in

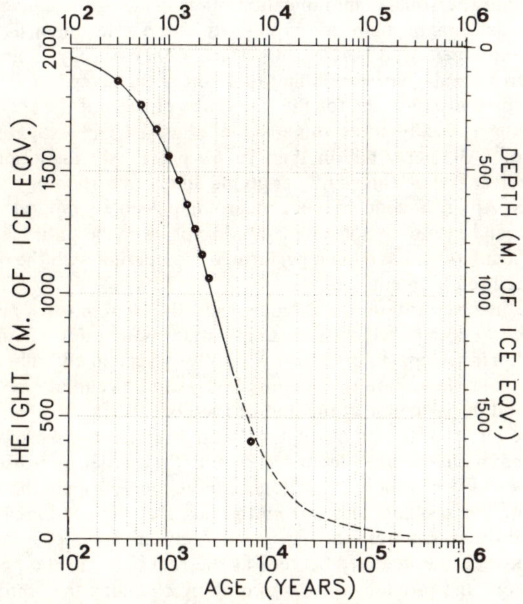

Fig. 8. Theoretical time scale for the Dye 3 deep ice core, compared with ages obtained by experimental dating (points marked by circles). The dashed lower part of the theoretical curve is less reliable than the upper part, mainly due to deviation of the model flow line from the actual one.

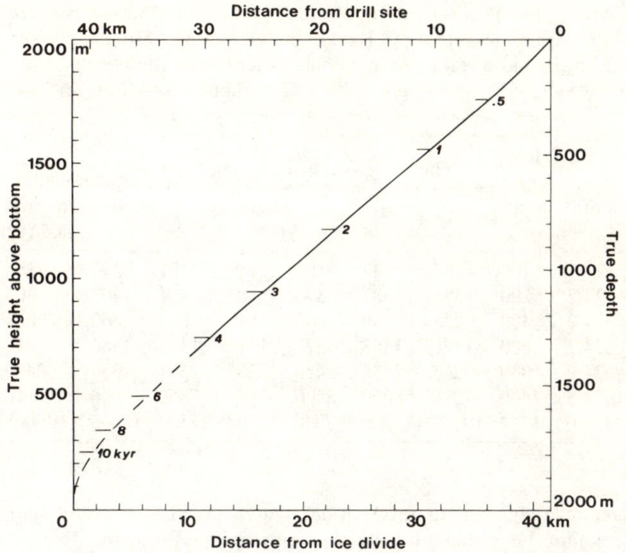

Fig. 9. Distance-depth diagram for the Dye 3 flow line obtained by flow model calculations. Ages in thousands of years based on experimental dating are indicated along the curve. The dashed lower part of the curve is less reliable than the upper part, mainly due to deviation of the model flow line from the actual one.

the deep ice [Herron et al., 1985] is in favor of the "soft-layer" profile. On the other hand, the other profile is in better agreement with the measured horizontal-surface velocity of 12.3 m/yr. Measurement of the horizontal-velocity profile in the deep drill hole would help to clarify this conflict.

The λ profiles shown in Figure 7b, which correspond to cases b_1 (dotted curve) and b_2 (dashed curve) described above, fit the measured profile as equally poorly as do the theoretical profiles shown in Figure 7a. It appears from the figure that the trends of accumulation rate and ice thickness affect the λ profile in opposite directions. Their combined effect on the Dye 3 λ profile is consequently small. At other ice sheet locations, however, the effect of one of the trends may dominate, or they may enhance each other and so contribute significantly to the shape of the λ profile.

Involvement of the "basal" undulations introduces a wavy shape to the λ profile (the dashed curve in Figure 7c). However, the wave amplitudes are small, and even though the theoretical curve seems to have more features in common with the measured profile than the previous ones, it still predicts neither the oscillations in the upper part nor the relatively thick layers from 1000–1600 m depth. As compared to the "Sandwich"-model reference profile (the chain-dotted curve), the λ profile has a bulge in the 600–1400 m depth interval. This is mainly due to the general ice-thickness distribution along the flow line, with smaller ice thicknesses in the middle section than at the ice divide and at the drill site. See Figure 2a.

Finally, considering also the irregular accumulation rate distribution, a considerable improvement is achieved as regards the prediction of the observed oscillations in the upper part of the λ profile, (Figure 7d). However, the relatively thick layers indicated by the line segments and the circle marks deeper in the core are still not explained.

As appears from the plot of the time scale shown in Figure 8, which corresponds to the λ profile of Figure 7d, the age of the thick layers ranges between 3000 and 7000 years. Even though it is tempting to interpret the thick layers as indications of increased

TABLE 1. Theoretical Time Scales for the Dye 3 Core Compared With Ages Obtained by Experimental Dating. Models denoted a, b_1, b_2, c, and d refer to the models described in the text. a_1 and a_2 are "Sandwich"—models, with and without a soft bottom layer, respectively.

True depth (m)	Ice eqv. depth (m)	Theoretical age according to model a_1	a_2	b_1 (yr)	b_2	c	d	Experimental age (yr)
123.7	100	187	187	187	187	191	191	185
224.2	200	385	387	390	384	401	403	404
424.2	400	822	830	840	818	857	860	861
624.2	600	1326	1348	1377	1319	1391	1386	1416
924.2	900	2268	2336	2420	2257	2344	2336	2400
1633.5	1609	7070	8020	8500	7560	7850	7850	(7110)
1786.0	1762	10100	12300	12900	11700	12500	12500	(10000)?

accumulation rate in this period, which comprises the climatic optimum, nevertheless caution is appropriate. From the distance/depth-age diagram shown in Figure 9 it can be seen that the layers in question were originally deposited between 2.5 and 17 km from the ice divide. As previously mentioned, this section of the model flow line deviates considerably from the real flow line, which has a more southerly course, passing an area of relatively thin ice (see Overgaard and Gundestrup, [1985, Figs. 2 and 5]). The layers originally deposited at such a location are subjected to a positive vertical strain-rate component, which is partly counteracting the general thinning of the layers, as they sink into the ice sheet. Therefore layers originating from locations with relatively thin ice will thin less than other layers. This effect may, partly or entirely, explain the thick layers encountered in the 1000–1600 m depth interval of the core.

The time scales corresponding to the λ profiles depicted in Figure 7 are illustrated in Table 1. The headings of columns a_1, a_2, b_1, b_2, c, and d refer to the sub-plot lettering of Figure 7 and to the model descriptions given above. Also shown in Table 1 are experimentally determined ages, obtained by counting seasonals of $\delta(^{18}O)$, dust, and acidity.

Down to a depth of 955 m of ice equivalent the experimental ages are based on a year by year dating and the corresponding dating errors are judged to be less than ± 3 years. At greater depths, the seasonals have been measured intermittently only, and consequently the age corresponding to the 1609 m of ice-equivalent depth, which is based on integration of a partly estimated λ profile, has a much greater uncertainty (probably less than a few hundred years, though). The age of 10,000 years attached to the depth of 1762 m of ice equivalent corresponds to the termination of the glaciation as indicated, e.g., by an abrupt change of the ^{18}O concentration in the Dye 3 core [Dansgaard et al., 1985].

Down to 600 m ice-equivalent depth the theoretical time scale corresponding to model d gives a rather close fit to the experimental one. At greater depths, however, discrepancies of 2000–3000 years occur between the theoretical ages, as well as between these and the experimental ones.

Discussion and Conclusions

The discussion of the various theoretical λ profiles depicted in Figure 7 has shown that in general all the features considered may have a significant influence on the λ profile and consequently also on the time scale. At a specific location one or more of the features may of course be indiscernible or they may counterbalance each other, as is the case with the trends of accumulation rate and ice thickness for the Dye 3 location (see Figure 7b). As can be seen in Figure 7 and Table 1, accurate predictions are obtained for the upper 600 m of the Dye 3 core only after introducing the perturbations of the base elevations and the perturbations of the accumulation rate. As regards the time scale, the basal perturbations are the more important (see Table 1), whereas the accumulation-rate perturbations account for the major improvement of the λ profile (see Figure 7d).

Unfortunately the significant impact on the λ profile of upstream accumulation-rate perturbations was not realized by the interpretation, in terms of past accumulation rates, of the λ profile measured along the 390 m long ice core, drilled at Dye 3 in 1971, [Reeh et al., 1978]. At that time, the wavy shape of the λ profile was interpreted as being due to temporal variations of the accumulation rate. The present investigation, however, has documented that the wavy shape should be ascribed to upstream short distance variations of the accumulation rate.

With increasing depth, the effect of the accumulation rate perturbations diminishes, their influence on the flow pattern being damped off (compare columns c and d in Table 1). This means that the rather speculative way of estimating the accumulation-rate perturbations for a major part of the flowline, as described in the general description of the ice-flow model, will not seriously affect the calculated λ and age profiles for the deep part of the Dye 3 core. As regards this part of the core, the investigation shows that the accumulation-rate trend, the detailed base-elevation profile, and the ice-fabric induced softening of the deep ice layers, are all the more important. Thus in order to improve the theoretical prediction of the λ and age profiles below, say, 1100 m depth, more accumulation-rate, ice-thickness, and base-elevation data are needed along the real turning towards the south flow line (see Figure 1).

Also, 3-dimensional effects may show up to be important. As regards the "basic" flow, for example, surface strain net measurements along a 24 km long section of the flow line upstream of Dye 3 by Whillans (personal communication) indicate a slightly converging flow, which might help to explain the apparent inconsistency between the measured horizontal surface velocity at Dye 3 and the predicted velocity, corresponding to a soft shear layer.

Also the consequences for the λ and age profiles of the ice sheet base being typically three dimensional should be investigated by means of a 3D-perturbation theory. As previously mentioned, a simplified study of this kind, applying linear ice rheology, is in progress. Any firm conclusions concerning possible temporal variations of past accumulation rates should not be made until ice flow model calculations have been performed, complying with the above mentioned requirements.

As to the interpretation of the deepest 200–300 m of the λ profile, the limitations imposed by the steady-state assumption should be stressed. This deepest part of the λ profile is very likely influenced by changing flow pattern, ice thickness and accumulation rate during and in particular at the end of the glaciation.

Moreover, the amplitude of the basal irregularities is so large compared with the ice thickness that a first order perturbation theory is insufficient below a distance above the bottom in the order of 300 m. The problem arises from the fact that the combined first- and zero-order solutions do not completely satisfy the boundary condition that the velocity be zero at the real base. In the present calculations the problem was overcome by changing the combined velocity fields below a distance of 200–300 m from the base so that this boundary condition is strictly fulfilled. The changed solution satisfies the equation of continuity. However, the dynamic equations are violated. A correct treatment of the problem would require application of a second-order-perturbation theory. Comparison of the changed velocity fields with those determined by second-order-perturbation theory for the case of linear viscous ice rheology (Dahl-Jensen, in preparation), shows good agreement.

Finally, as regards the very deepest (say 90 m) portion of the ice core, it is questionable whether the layer sequence is at all continuous. Coincident sharp changes of different parameters measured along the core can probably only be explained as being due to breaks (shear planes) in the core related to abrupt gradient changes in the medium or small scale bedrock profile.

Acknowledgments. The computer expenses for developing and running the numerical ice flow model were partly funded by the Danish Natural Science Research Council.

References

Budd, W.F., The dynamics of ice masses, *ANARE scientific reports, Series A (IV) Glaciology, Publication Number 108*, Antarctic Division, Department of Supply, Melbourne, 1969.

Budd, W.F., Ice flow over bedrock perturbations, *Journal of Glaciology, 9* (55), 29–48, 1970.

Dansgaard, W., and S.J. Johnsen, A flow model and a time scale for the ice core from Camp Century, Greenland, *Journal of Glaciology, 8* (53), 215–223, 1969.

Dansgaard, W., H.B. Clausen, N. Gundestrup, S.J. Johnsen, and C. Rygner, Dating and climatic intepretation of the two deep Greenland ice cores, this volume, 1985.

Drew, A.J.R., Glacial movements in Greenland from doppler satellite observations, M.S. thesis, The Ohio State University, Columbus, 1982.

Gow, A.J., and R. Rowland, On the relationship of snow accumulation to surface topography at "Byrd Station", Antarctica, *Journal of Glaciology, 5* (42), 843–847, 1965.

Haefeli, R., A numerical and experimental method for determining ice motion in central parts of ice sheets, General Assembly of Berkeley, Commission of Snow and Ice (Proceedings of the General Assembly of Berkeley, 1963); *IAHS-AISH Publication No. 61*, 253–260, 1963.

Hammer, C.U., H.B. Clausen, W. Dansgaard, N. Gundestrup, S.J. Johnsen, and N. Reeh, Dating of Greenland ice cores by flow models, isotopes, volcanic debris, and continental dust, *Journal of Glaciology, 20* (82), 3–26, 1978.

Hammer, C.U., Acidity of polar ice cores in relation to absolute dating, past volcanism, and radio-echoes, *Journal of Glaciology, 25* (93), 359–372, 1980.

Herron, S.L., C.C. Langway, Jr., and K.A. Brugger, Ultrasonic velocities and crystalline anisotropy in the ice core from Dye 3, Greenland, this volume, 1985.

Hutter, K., F. Legerer, and U. Spring, First-order stresses and deformations in glaciers and ice sheets, *Journal of Glaciology, 27* (96), 227–270, 1981.

Johnsen, S.J., Stable isotope homogenization of polar firn and ice, Symposium on Isotopes and Impurities in Snow and Ice (Proceedings of the Grenoble Symposium, 1975), *IAHS-AISH Publication No. 118*, 210–219, 1977.

Mock, S., Geodetic positions of bore-hole sites of the Greenland Ice Sheet Program, *Report 76-41*, U.S. Army Corps of Engineers, Cold Regions Research and Engineering Laboratory, Hanover, N.H., 1976.

Nye, J.F., The motion of ice sheets and glaciers, *Journal of Glaciology, 3* (26), 493–507, 1959.

Overgaard, S., and N.S. Gundestrup, Bedrock topography of the Greenland Ice Sheet in the Dye 3 area, this volume, 1985.

Paterson, W.S.B., *The Physics of Glaciers*, Second Edition, Pergamon Press Ltd., New York, 31, 1981.

Paterson, W.S.B., and W.F. Budd, Flow parameters for ice sheet modelling, *Cold Regions Science and Technology*, *6*, 175–177, 1982.

Philbert, K., and B. Federer, On the temperature profile and the age profile in the central part of cold ice sheets, *Journal of Glaciology, 10* (58), 3–14, 1971.

Ragle, R.H., and T.C. Davis, South Greenland traverses, *Journal of Glaciology, 4* (31), 129–131, 1962.

Raynaud, D., C. Lorius, W.F. Budd, and N.W. Young, Ice flow along an I.A.G.P. flow line and interpretation of data from an ice core in Terre Adélie, Antarctica, Symposium on Dynamics of Large Ice Masses, Ottawa, August 1978, *Journal of Glaciology, 24* (90), 103–115, 1979.

Reeh, N., H.B. Clausen, W. Dansgaard, N. Gundestrup, C.U. Hammer, and S.J. Johnsen, Secular trends of accumulation rates at three Greenland stations, *Journal of Glaciology, 20* (82), 27–30, 1978.

Shoji, H., and C.C. Langway, Jr., Mechanical properties of fresh ice cores from Dye 3, Greenland, this volume, 1985.

Whillans, I.M., Ice flow along the Byrd station strain net work, Antarctica, Symposium on Dynamics of Large Ice Masses, Ottawa, August 1978, *Journal of Glaciology, 24* (90), 15–26, 1979.

Whillans, I.M., and S.J. Johnsen, Longitudinal variations in glacial flow: Theory and test using data from the Byrd Station strain network, Antarctica, *Journal of Glaciology*, *29*, 78–97, 1983.

Zwally, H.J., R.A. Bindschadler, A.C. Brenner, T.V. Martin, and R.H. Thomas, Surface elevation contours of Greenland and Antarctic ice sheets, *Journal of Geophysical Research, 88* (C3), 1589–1596, 1983.

^{10}Be VARIATIONS IN POLAR ICE CORES

J. Beer, M. Andrée, H. Oeschger, and B. Stauffer

Physics Institute, University of Berne, Switzerland

R. Balzer, G. Bonani, Ch. Stoller, M. Suter, and W. Wölfli

Laboratorium für Kernphysik, Swiss Federal Institute of Technology, Zürich, Switzerland

R. C. Finkel[1]

University of California, La Jolla, U.S.A.

Abstract. ^{10}Be is produced in the atmosphere by cosmic radiation and has a half life of $1.5 \cdot 10^6$ years. The produced ^{10}Be atoms are attached to aerosols and washed out by precipitation. The mean residence time in the atmosphere is about one year. Annual snow layers with their ^{10}Be contents are ideally preserved in polar ice sheets.

With the new accelerator based mass-spectrometry technique it is possible to measure the natural ^{10}Be concentrations in ice samples of 1 kg each. ^{10}Be, with its short residence time and long half life, is very well suited for the study of variations of the interaction of cosmic radiation with the atmosphere. We measured firn samples from Dye 3 from the surface to about 100 m depth, representing the precipitation of the last 140 years. One sample per 2 years has been measured. The results are compared with the sunspot-number record. A clear indication of the 11 year cycle modulation has been obtained.

In addition the ^{10}Be activities of ice samples from 1600 m–1900 m depth have been measured. Samples from this depth interval represent precipitation from about 13,000 yrs B.P. to 6000 yrs BP. The results are compared with results from Antarctica, published by Raisbeck. By measuring the $^{10}Be/^{36}Cl$ ratio, it should be possible to date ice older than about 50,000 yrs.

Introduction

^{10}Be ($T_{1/2} = 1.5 \cdot 10^6$ y) is the second most abundant long-lived radioisotope after ^{14}C. Both isotopes are produced by the interaction of the galactic cosmic radiation with the atmosphere. About 70% of the ^{10}Be and ^{14}C production takes place in the stratosphere. ^{10}Be is mainly produced by spallation reactions on nitrogen and oxygen, ^{14}C by the ^{14}N (n,p) ^{14}C reaction. ^{10}Be becomes attached to aerosols within a very short time. If this process takes place in the troposphere, ^{10}Be is deposited rapidly on the earth's surface by precipitation. If ^{10}Be is produced in the stratosphere, some latitudinal mixing occurs. Most of the ^{10}Be is transferred to the troposphere during spring and early summer when mainly at median latitudes large stratospheric air masses enter the troposphere. The fallout pattern therefore does not reflect the increased production rate in the polar stratosphere. The mean residence time of ^{10}Be in the atmosphere is about one to two years. ^{10}Be removed from the atmosphere by precipitation is either preserved in snow and ice layers, in the topsoil and the biosphere or it enters the hydrosphere (oceans and lakes) where it is transported to the sediments.

The ^{14}C atoms produced in the atmosphere are oxidized and homogeneously distributed in the atmospheric CO_2 system from where they are exchanged by molecular processes with the ocean and the biosphere. The atmospheric residence time is about five years. Many precise ^{14}C measurements on tree rings covering the last 8000 years have been carried out during the last years [Suess, 1980; Stuiver and Quay, 1980; De Jong and Mook, 1980; Bruns et al., 1980; Pearson, 1980]. The results clearly show that the atmospheric ^{14}C concentration has not been constant. Possible causes of these variations with amplitudes of 1 to 2% are:

–changes in the production rate due to changes in the galactic cosmic ray intensity. The incoming cosmic ray flux is modulated by the magnetic properties of the solar wind plasma and the intensity of the geomagnetic field.

—changes in the global carbon cycle [Siegenthaler et al., 1980]. The reservoir sizes and exchange fluxes can be influenced by changes in the environmental conditions.

Fluctuations in the ^{14}C production rate are strongly dampened due to the large size of the atmospheric CO_2 reservoir and the exchange processes with the ocean. Model calculations show that, e.g., the production variation induced by the 11-year solar cycle is attenuated by a factor of 100. Because of its rapid transfer from the atmosphere to the geosphere, ^{10}Be responds with a much greater amplitude to changes of the production rate. The 11-year production variation is attenuated only by about 20%. Until 1977 only a few ^{10}Be measurements were made, mainly on ocean sediments [Somayajulu, 1977; Finkel et al., 1977]. Because of its ~100 times smaller global production rate compared to ^{14}C and its ~300 times larger half life, the detection of ^{10}Be by conventional low level counting techniques is very difficult. For the first ^{10}Be measurement on polar ice which represents the best record of precipitation, $1.2 \cdot 10^6$ kg of meltwater were processed [McCorkell et al., 1967]. Since the development of the accelerator mass spectrometer in 1977, samples of about 1 kg containing $\sim 10^7$ ^{10}Be atoms are sufficient for a measurement. The first sets of ^{10}Be measurements in ice cores from Antarctica using the accelerator technique show very promising results [Raisbeck et

[1] Now at: Livermore National Laboratory, Livermore, California.

TABLE 1. T- and ^{137}Cs-Activities in the Firn Core From Dye 3, Greenland

Year	T-activity (TU)	^{137}Cs-activity (dph/kg)
1950	0	22 ± 22
1951	0	
1952	0	55 ± 32
1953	13 ± 3	
1954	100 ± 20	165 ± 25
1955	32 ± 8	
1956	161 ± 26	104 ± 37
1957	129 ± 27	
1958	401 ± 44	212 ± 28
1959	530 ± 42	
1960	158 ± 18	94 ± 24
1961	240 ± 45	
1962	1185 ± 47	640 ± 53
1963	3940 ± 140	
1964	3200 ± 190	446 ± 36
1965	780 ± 80	
1966	505 ± 40	69 ± 25
1967	457 ± 32	
1968	238 ± 37	35 ± 19
1969	313 ± 39	
1970	209 ± 23	55 ± 17
1971	80 ± 15	
1972	117 ± 14	18 ± 13
1973	59 ± 8	
1974	101 ± 9	20 ± 12
1975	60 ± 11	
1976	52 ± 8	12 ± 12
1977	71 ± 12	
1978	90 ± 10	10 ± 10
1979	33 ± 8	

al., 1981]. To study ^{10}Be variations in ice cores for at least the last 10^5 years we started with two sets of samples drilled at Dye 3, Greenland. The main goal of set A for 1900 to 1976 is to study short-term fluctuations caused by changes of the solar activity, i.e. the 11-year solar cycle. Set B contains 14 samples distributed over the depth range 1300 to 1950 m corresponding to a time period from 3600 BP to ~30,000 BP. They should provide some information on long-term fluctuations and on the changing conditions during the transition from the glacial to the post glacial time (10,000–13,000 BP).

Sample Preparation

Samples were prepared from two ice cores drilled at Dye 3, Greenland (65°11′N, 43°50′W), in the frame of GISP (Greenland Ice Sheet Project) realized by an American-Danish-Swiss collaboration. The samples of set A were prepared from a 70 m shallow firn core. The $\delta^{18}O$ profile (approximately eight samples per year) measured by the Danish group was used to cut the core into pieces containing the precipitation of one year. Each sample was mechanically cleaned and melted. Aliquots for tritium and chemical analysis were taken. Before adding Be- (1.18 mg Be) and Cl-carrier (2 mg Cl) the precipitation of two years was combined into one sample to increase the ^{10}Be concentration. First the water volume (2–6 kg, cf. Table 2) was reduced by evaporation to about 30 ml. Then the chlorine was separated by precipitation of AgCl and purified as described elsewhere [Nishiizumi et al., 1979]. After an additional volume reduction to 2 ml the samples were analyzed by gamma spectroscopy. Beryllium acetylacetonate in the presence of EDTA was extracted into $CHCl_3$ and evaporated after adding HCl. The organic material was oxidized with aqua-regia. $Be(OH)_2$ was precipitated with NH_4OH and converted to BeO by ignition at 950°C in a quartz crucible. The 14 samples of set B were prepared from ice of the 2037 m-long deep core, reaching bedrock. Depending on the depth, one sample represents the precipitation of about 10 to 250 years. The samples were processed in the same way as the samples of set A.

Measurements

The ^{10}Be concentrations were measured using the EN-tandem accelerator facility of the ETH Zürich. This system was designed in 1978 to detect, in a first step, ^{14}C and ^{10}Be, and in a second step, ^{36}Cl and ^{26}Al in natural samples. The Cs sputter ion source produced BeO currents of up to 1 μA leading to count rates of up to 10^3 cph for a typical $^{10}Be/^{9}Be$ ratio of $5 \cdot 10^{-13}$. The background is of the order of 10^{-14} depending on the boron content of the sample. The accelerator mass spectrometer is described in more detail elsewhere [Wölfli et al., 1983].

The gamma activity of the samples was measured with a 64 cc Ge(Li) detector. The tritium content was determined by a commer-

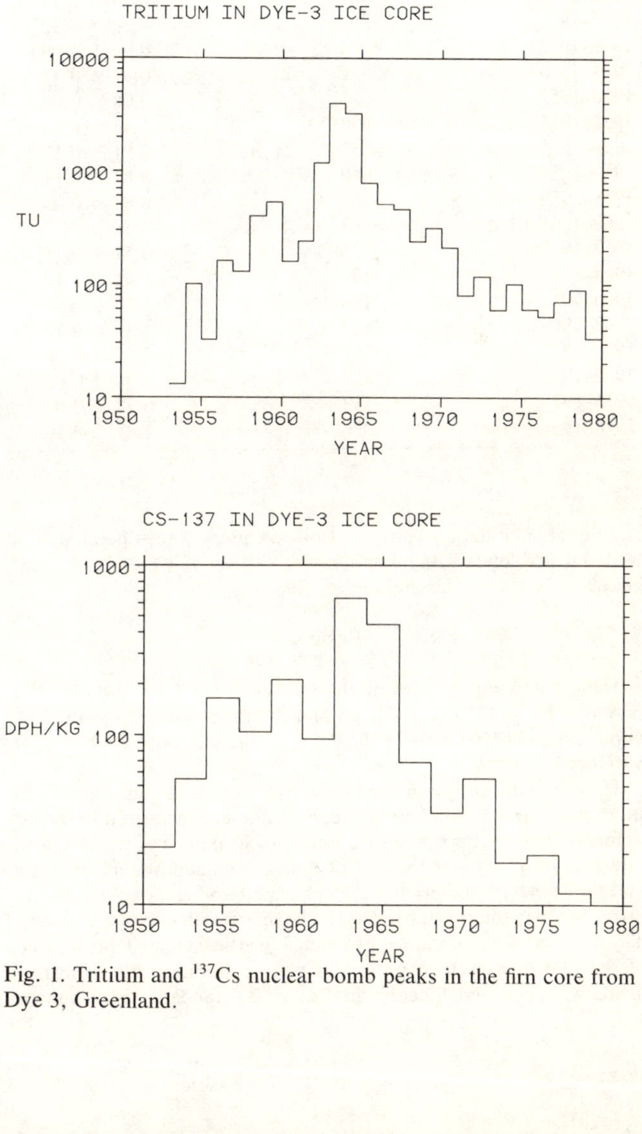

Fig. 1. Tritium and ^{137}Cs nuclear bomb peaks in the firn core from Dye 3, Greenland.

TABLE 2. ^{10}Be Concentrations in the Dye 3 Firn Core

Year	Sample weight (g)	^{10}Be concentration 10^4 atoms/g
1900–1901	3853	1.11 ± 0.05
1902–1903	3192	0.91 ± 0.05
1904–1905	3354	1.09 ± 0.06
1906–1907	2295	1.27 ± 0.07
1908–1909	4236	0.79 ± 0.05
1910–1911	2970	0.81 ± 0.06
1912–1913	4623	0.87 ± 0.05
1914–1915	3558	1.10 ± 0.10
1916–1917	3460	0.85 ± 0.05
1918–1919	3584	0.77 ± 0.09
1920–1921	3255	0.98 ± 0.14
1922–1923	3892	0.74 ± 0.09
1924–1925	3028	1.01 ± 0.20
1926–1927	3853	0.85 ± 0.14
1928–1929	4604	0.57 ± 0.07
1930–1931	2836	0.98 ± 0.13
1932–1933	3230	0.97 ± 0.09
1934–1935	3392	0.86 ± 0.10
1936–1937	2941	0.97 ± 0.13
1938–1939	6559	0.72 ± 0.05
1940–1941	4493	
1942–1943	3401	0.74 ± 0.11
1944–1945	2821	0.80 ± 0.12
1946–1947	4274	0.90 ± 0.08
1948–1949	2387	0.93 ± 0.11
1950–1951	2673	0.98 ± 0.08
1952–1953	2538	1.40 ± 0.12
1954–1955	3652	0.81 ± 0.12
1956–1957	2660	0.83 ± 0.11
1958–1959	3560	0.62 ± 0.06
1960–1961	3573	
1962–1963	2256	0.57 ± 0.15
1964–1965	3199	1.41 ± 0.12
1966–1967	2140	1.17 ± 0.17
1968–1969	2843	1.10 ± 0.10
1970–1971	3105	0.93 ± 0.08
1972–1973	3629	0.78 ± 0.07
1974–1975	3118	0.78 ± 0.07
1976–1977	3026	1.47 ± 0.10

cial liquid scintillation counter. Both counters were operated in a well-shielded underground laboratory, 70 m of water equivalent below the surface [Oeschger et al., 1981].

Results

Table 1 and Figure 1 show the tritium and ^{137}Cs results for the period 1950 to 1979. The shape and the maximum of the nuclear bomb pulses in 1963 confirm the dating of the ice core based on $\delta^{18}O$ variations.

The ^{10}Be concentrations and the weights of set A and B are given in Tables 2 and 3. All samples (except 2) were measured twice, at different times. The measuring time (20–30 min) was divided into intervals of 50 sec. At the end of a measurements the mean value and the standard deviation were calculated and compared with the error due to counting statistics. The larger of both errors was taken. The final result was obtained by calculating the weighted mean value of the two measurements. For absolute calibration the measurements were periodically compared to a ^{10}Be standard with a known ^{10}Be/^{9}Be ratio. The boron present in the Be beam was stopped and measured in an absorber chamber. Using a ^{10}B beam it was possible to determine the ^{10}Be background induced by nuclear reactions of ^{10}B in the absorber chamber [Middleton et al., 1981]. This information was used to correct the data.

In Figure 2 the data of set A are plotted together with the negative sunspot numbers. The comparison of the spline function fit of our data with the sunspot curve shows good agreement. The maximum of the spectral density function is at 13 ± 3 years. The cross correlation between the Be data and the sunspot numbers yields a phase lag of the ^{10}Be variations of 1.5 years that is consistent with the mean atmospheric residence time of ^{10}Be. Model calculations of the variation of the ^{10}Be production rate induced by the 11-year solar cycle predict changes of about 60% [Oeschger et al., 1970] in agreement with the variations shown in Figure 2. Between 1900 and 1960 the ^{10}Be concentration shows a slight decreasing trend which could be explained by the observed slow rise of the solar activity [Eddy, 1976]. After 1960 the ^{10}Be concentration rises again with greater amplitudes. There is no clear indication of a bomb peak due to nuclear weapon tests as observed for ^{137}Cs, T (Figure 1), and ^{36}Cl [Elmore et al., 1982]. The mean value between 1900 and 1976 is (0.93 ± 0.22) 10^4 atoms per gram of ice.

Table 3 and Fig. 3 give the results of the samples from the deep core. Because there is no general agreement about the age below 1780 m (corresponding to 10,000 BP) the data are listed as a function of depth. The time scale added in Figure 3 is based on an ice flow model [Hammer et al., 1978]. The ^{10}Be concentration changes dramatically between 1810 and 1830 m corresponding to the end of the last ice age (~ 10,000 BP). The mean value of the first six points above 1780 m is (0.97 ± 0.25) 10^4 atoms/g concurring with the value found for the twentieth century ((0.93 ± 0.22) 10^4 atoms/g).

Discussion

The ^{10}Be concentration of the time period 1900 to 1976 reflects to some extent the 11 year solar cycle. This conclusion is supported by the size of the amplitude, the spectral density function, and the phase lag. Since measurements of ^{10}Be concentration in monthly rainwater in France [Raisbeck et al., 1979a] show variations of a factor of 3, it is not surprising to find some "meterological noise" in the two year mean values. There are even some indications based on measurements of the atmospheric ^{7}Be concentration that the frequency of stratospheric air injections into the troposphere is triggered by solar events [Reiter, 1975]. The easiest way to eliminate

TABLE 3. ^{10}Be Concentrations in the Dye 3 Deep Core

Depth (m)	Sample weight (g)	^{10}Be concentration (10^4 atoms/g)
1314–1315	1856	0.59 ± 0.10
1397–1398	1701	1.16 ± 0.14
1517–1517.5/1537–1537.5	1932	1.20 ± 0.15
1636–1636.5/1657–1657.5	1842	0.72 ± 0.13
1714.5–1715.5	2175	1.05 ± 0.12
1775.5–1776.5	1837	1.11 ± 0.12
1800.5–1801.5	1687	1.60 ± 0.15
1810.5–1811.5	1781	0.86 ± 0.09
1832.5–1833.5	1200	2.45 ± 0.13
1852.5–1853.5	1445	2.87 ± 0.14
1873.5–1874.5	1804	2.43 ± 0.16
1895.5–1896.5	1272	1.43 ± 0.11
1913.5–1914.5	1689	1.93 ± 0.12
1930–1931	2004	1.52 ± 0.15

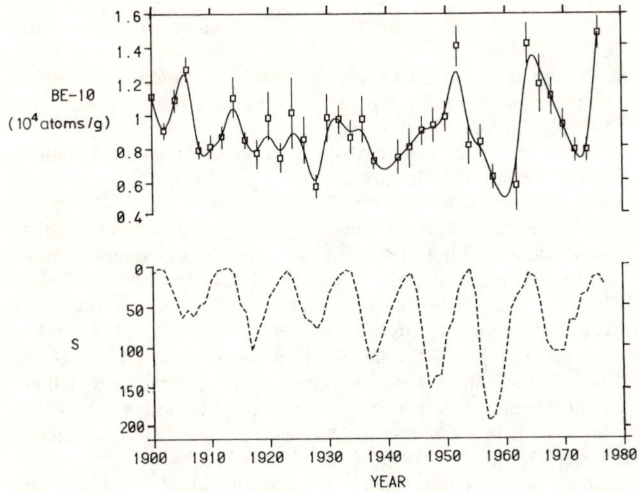

Fig. 2. Comparison of ^{10}Be concentrations (two years means) in the Dye 3 firn core with the negative sunspot numbers.

this meteorological noise would be to normalize the ^{10}Be data to a tracer with a relatively constant stratospheric concentration and a similar deposition mechanism. The absence of a nuclear bomb peak is expected because the environmental ^{9}Be concentration and the n-activation cross section are small.

Using the fallout pattern of Lal and Peters [1967] and the average rainfall for the latitude of Dye 3 [Møller, 1951] the mean ^{10}Be concentration of (0.93 ± 0.22) 10^4 atoms/g corresponds to a global deposition rate of 0.016 $cm^{-2}sec^{-1}$ which agrees well with the value 0.018 $cm^{-2}sec^{-1}$ published by Amin [Amin et al., 1975] and 0.020 $cm^{-2}sec^{-1}$ measured in an ice sample from Camp Century, Greenland [McCorkell et al., 1967] but is significantly lower than the values given by Raisbeck et al. of 0.042 $cm^{-2}sec^{-1}$ and 0.10 $cm^{-2}sec^{-1}$ deduced from rainwater [Raisbeck et al., 1979a] and marine sediments [Raisbeck et al., 1979b] respectively.

There are three possible explanations for the significant change of the ^{10}Be concentration at the transition from Wisconsin to Holocene. The increase by a factor 2.5 to 3 can be caused by an enhanced production rate, by changes in the atmospheric circulation and mixing processes or by changes in the precipitation rates. Other parameters like δ^{18}O, CO_2 content, anion and dust concentration that have been measured on the same ice core [Dansgaard et al., 1985; Stauffer et al., 1985; Herron and Langway, 1985; Hammer et al., 1984] also show significant variations at this depth. In Figure 4 the ^{10}Be values around the transition are plotted together with Danish δ^{18}O data. It is surprising how well the δ^{18}O curve fits the ^{10}Be points (except the first). Raisbeck et al. [1981] found the same features in an ice core from Dome C, Antarctica; an increase by a factor of 2 to 3 between 10,000 and 15,000 BP and a good correlation with the δ^{18}O curve. Considering that the increase of the ^{10}Be concentration coincides with a strong climatic change, it is highly probable that this effect can be attributed to serious changes of the atmospheric circulations and precipitation rates. Herron and Langway [1985] found that, based on sulfate measurements, precipitation at Dye 3 during the last ice age was lower by a factor of 2 to 3. However, there are indications that the deposition rate in Antarctica was rather constant during this time. With the present information it is difficult to decide if the production rate of ^{10}Be was higher during the Wisconsin. Figure 1 and results deduced from ice samples of the Maunder minimum [Raisbeck et al., 1981] show that the production

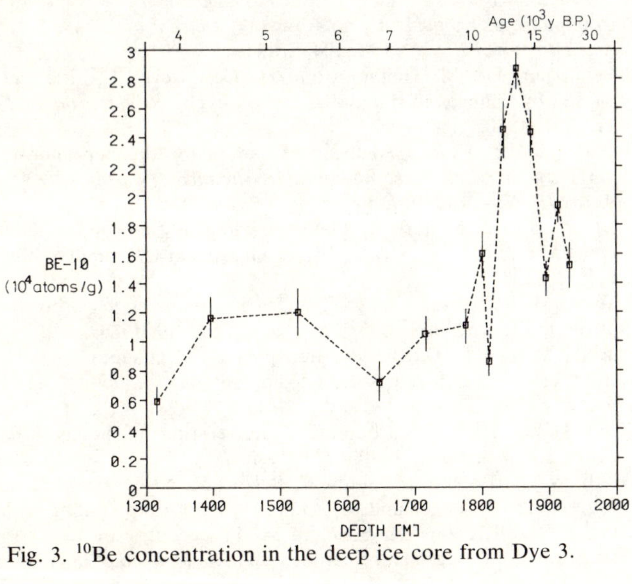

Fig. 3. ^{10}Be concentration in the deep ice core from Dye 3.

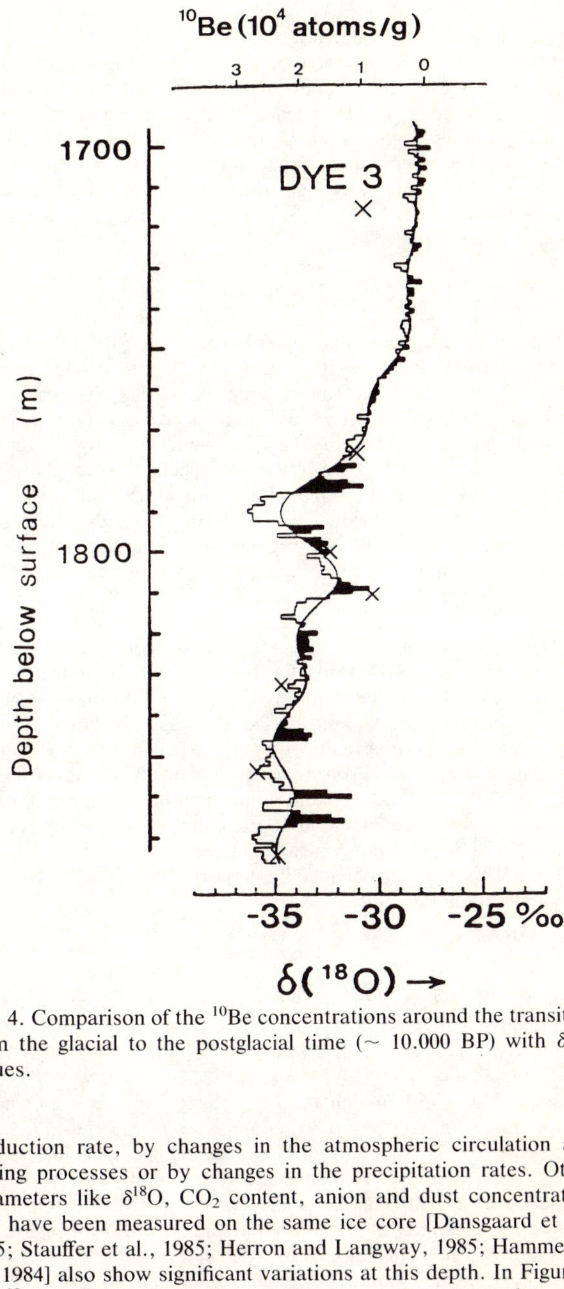

Fig. 4. Comparison of the ^{10}Be concentrations around the transition from the glacial to the postglacial time (~ 10.000 BP) with δ^{18}O values.

rate during periods of very low solar activity is increased by less than 100%. Thus it seems improbable that changing solar magnetic properties alone could account for the observed threefold higher ^{10}Be concentrations. The Holocene ^{10}Be concentration at Dome C is about five times higher than at Dye 3. This difference can be explained by the very low accumulation rate of about 4 cm at Dome C compared to 50 cm at Dye 3 and the fact that the fallout decreases between the latitudes 65° (Dye 3) and 75° (Dome C) by a factor of 2 [Lal and Peters, 1967].

Conclusions

Despite some meteorologic disturbances, the ^{10}Be data seem to reflect solar activity as well as climatic changes. If confirmed by additional measurements from different sites, this observation has important implications. Records of solar activity provide basic information for understanding the solar cycle mechanism. Comparisons of ^{10}Be and ^{14}C data sequences enable us to distinguish variations due to fluctuations of the global carbon cycle. The correlation of the ^{10}Be concentration with δ^{18}O values and other climatic parameters could be very helpful in studying not only the climate of the last 10^5 to 10^6 years but also the extent to which climatic changes are influenced by the sun.

Acknowledgements. We thank A. Salem for help in preparing the BeO samples, U. Schotterer for the preparation of the tritium samples and M. Heimann for assistance in the analysis of the results. Further thanks are due to the U.S. National Science Foundation for the continuing support, especially for the logistic help during the field season at Dye 3 and to the Commission for Scientific Research in Greenland supporting the development of the Danish drill. We are indebted to all members of the American-Danish-Swiss groups who contributed to the successful deep drilling at Dye 3. This work was financially supported by the Swiss National Science Foundation.

References

Amin, B. S., D. Lal, B. L. K. Somayajulu, Chronology of marine sediments using the ^{10}Be method: intercomparison with other methods, *Geochemica et Cosmochemica Acta, 39,* 1187–1192, 1975.

Bruns, M., K. O. Münnich and B. Becker, Natural radiocarbon variations from AD 200 to 800, *Radiocarbon, 22,* 273–277, 1980.

Dansgaard, W., H. B. Clausen, N. Gundestrup, S. J. Johnsen, and C. Rygner, Dating and climatic interpretation of two deep Greenland ice cores, this volume, 1985.

DeJong, A. F. M. and W. G. Mook, Medium-term atmospheric ^{14}C variations, *Radiocarbon, 22,* 267–272, 1980.

Eddy, J. A., The Maunder Minimum, *Science, 192,* 1189–1202, 1976.

Elmore, D., L. E. Tubbs, D. Newman, X. Z. Ma, R. Finkel, K. Nishiizumi, J. Beer, H. Oeschger and M. Andrée, The ^{36}Cl bomb pulse measured in a 100 m ice core from Dye 3, Greenland, *Nature, 300,* 735, 1982.

Finkel, R., S. Krishnaswami and D. L. Clarck, ^{10}Be in arctic ocean sediments, *Earth and Planetary Science Letters, 35,* 199–204, 1977.

Hammer, C. U., H. B. Clausen, W. Dansgaard, N. Gundestrup, S. J. Johnsen and N. Reeh, Dating of Greenland ice cores by flow models, isotopes, volcanic debris and continental dust, *Journal of Glaciology, 20,* 3–26, 1978.

Hammer, C. U., H. B. Clausen, W. Dansgaard, A. Neftel, P. Kristinsdottir, and E. Johnson, Continuous impurity analysis along the Dye 3 deep core, this volume, 1984.

Herron, M. M. and C. C. Langway, Jr., Chloride, nitrate and sulfate in the Dye 3 and Camp Century, Greenland, ice cores, this volume, 1984.

Lal, D. and B. Peters, Cosmic ray produced radioactivity on the earth, in *Handbuch der Physik,* Vol. 462, edited by S. Flügge, pp. 551–612, Springer, Berlin, 1967.

McCorkell, R., E. L. Fireman and C. C. Langway, Jr., Aluminum-26 and Beryllium-10 in Greenland ice, *Science, 158,* 1690–1692, 1967.

Middleton, R., J. Klein and Tang Hongquing, Instrumentation of an EN tandem for the detection of ^{10}Be, Proc. Symposium on Accelerator Mass Spectrometry, pp. 57–86, Argonne, 1981.

Moller, F., Vierteljahreskarten des Niederschlages für die ganze Erde, *Petermanns Geographische Mitteilungen, 95,* 1–7, 1951.

Nishiizumi, K., J. R. Arnold, D. Elmore, R. D. Ferraro, H. E. Gove, R. C. Finkel, R. O. Beukens, K. H. Chang, L. R. Kilius, Measurements of ^{36}Cl in Antarctic meteorites and Antarctic ice using a van de Graff accelerator, *Earth and Planetary Science Letters, 45,* 285–292, 1979.

Oeschger, H., J. Houtermans, H. Loosli and M. Wahlen, The constancy of cosmic radiation from isotope studies in meteorites and on the Earth, in Olsson, I. U., ed., Radiocarbon variations and absolute chronology, Proceedings of the 12th Nobel Symposium, 471–498, John Wiley and Sons, New York, 1970.

Oeschger, H., J. Beer, H. H. Loosli and U. Schotterer, Low level counting systems in deep underground laboratories, in Methods of Low-Level-Counting and Spectrometry, Proceedings of a Berlin Symposium, International Atomic Energy Agency, Vienna, Austria, pp. 459–474, 1981.

Pearson, G. W., High precision radiocarbon dating by liquid scintillation counting applied to radiocarbon time-scale calibration, *Radiocarbon, 22,* 337–345, 1980.

Raisbeck, G. M., F. Yiou, M. Fruneau, J. M. Loiseaux, M. Lieuvin and J. C. Ravel, Cosmogenic ^{10}Be: mean deposition rate and seasonal concentration variations in precipitation, *Nature, 282,* 279–280, 1979a.

Raisbeck, G. M., F. Yiou, M. Fruneau, J. M. Loiseaux, M. Lieuvin, J. C. Ravel and J. D. Hays, A search in a marine sediment core for ^{10}Be concentration variations during a geomagnetic field reversal, *Geophysical Research Letters, 6,* 717–719, 1979b.

Raisbeck, G. M., F. Yiou, M. Lieuvin, J. C. Ravel, M. Fruneau and J. M. Loiseaux, ^{10}Be in the environment: some recent results and their applications, *Proceedings of Symposium on Accelerator Mass Spectrometry,* 228–243, Argonne, 1981.

Reiter, R., Increased frequency of stratospheric injections into the tropopause as triggered by solar events, *Journal of Atmospheric and Terrestrial Physics, 38,* 504–510, 1975.

Siegenthaler, U., M. Heimann and H. Oeschger, ^{14}C variations caused by changes in the global carbon cycle, *Radiocarbon, 22,* 177–191, 1980.

Somayajulu, B. L. K., Analysis of causes for the beryllium-10 variations in deep sea sediments, *Geochemica et Cosmochemica Acta, 41,* 909–913, 1977.

Stauffer, B., A. Neftel, H. Oeschger, and J. Schwander, CO_2 concentration in air extracted from Greenland ice samples, this volume, 1985.

Stuiver, M. and P. D. Quay, Changes in atmospheric carbon-14 attributed to a variable sun, *Science, 207,* 11–19, 1980.

Stuiver, M. and P. D. Quay, Atmospheric ^{14}C changes resulting from fossil fuel CO_2 release and cosmic ray flux variability, *Earth and Planetery Science Letters, 53,* 349–362, 1981.

Suess, H. E., The radiocarbon record in tree rings of the last 8000 years, *Radiocarbon, 22,* 200–209, 1980.

Wölfli, W., G. Bonani, M. Suter, R. Balzer, M. Nessi, Ch. Stoller, J. Beer, H. Oeschger and M. Andrée, Radioisotope dating with the ETHZ EN tandem accelerator, Proceedings of the 11th Radiocarbon Conference, Seattle, 1982, *Radiocarbon, 25,* 1983.

DATING AND CLIMATIC INTERPRETATION OF TWO DEEP GREENLAND ICE CORES

W. Dansgaard, H. B. Clausen, N. Gundestrup, S. J. Johnsen[1], and C. Rygner

Geophysical Isotope Laboratory, University of Copenhagen, Denmark

Abstract. A continuous $\delta(^{18}O)$ profile is measured along the new deep ice core from Dye 3, South Greenland. It allows absolute dating back to at least 7000 yrs BP. The Pleistocene to Holocene δ shift is discussed. The Wisconsin ice in the deepest 300 m has so many δ features in common with the Camp Century (CC) ice core that they must be of climatic significance. The deep parts of the two δ profiles are indirectly dated by tentative correlation with a deep sea foraminifera record. The CC record seems to be continuous to more than 125,000 yrs BP, and it suggests a dramatic termination of the Eem/Sangamon interglacial. The Dye 3 record, however, is hardly continuous to more than 90,000 yrs BP, although the deepest 22 m of silty ice is probably of Eem/Sangamon origin. The basic criteria for an ideal deep drilling site are listed.

Introduction

One of the major objectives of the Danish participation in the Greenland Ice Sheet Program (GISP) was to measure the isotopic composition of firn and ice cores. More than 100,000 oxygen-18 analyses have been performed with the aims of absolute dating and extraction of the climatic signals that might be contained in long time series of oxygen-18 concentration, given in terms of δ, i.e. the relative per mille deviation from the oxygen-18 concentration in Standard Mean Ocean Water.

Summer snow is generally higher in δ than is winter snow, as already shown by Epstein and Benson [1959]. Figure 1 shows the deepest part of a continuous δ profile along the 400-m long ice core from Milcent, mid-Greenland (70.3°N; 44.6°W). δ is plotted as a function of depth corrected for vertical strain since the deposition. The dating shown to the right was obtained by counting summer peaks downward from the surface. This is how the δ cycles look, when the deposition conditions are ideal, i.e. no melting, and high accumulation rate, in this case 53 cm of ice equivalent per year.

In the Dye 3 area (65.2°N; 43.8°W), the accumulation rate is nearly the same, but there is some summer melting, which sometimes deforms the δ cycles. Nevertheless, they are interpretable (cp. Figure 1 in Dansgaard et al., [1982]), in particular if we cross-check with profiles of other seasonally varying parameters, such as concentrations of dust and other impurities [Langway, 1970; Hammer et al., 1978; Risbo et al., 1981; M. Herron, unpublished data], or with elevated acidity levels due to fall-out of volcanic acids observed shortly after great, well-known volcanic eruptions [Hammer, 1980], for example the 1783 Laki eruption in Iceland. So far, the last 3000 annual layers have been identified, and the absolute dating will be extended more than 7000 years back in time. As long as the mean accumulation rate is higher than 25 cm/yr, this technique allows absolute dating with an accuracy that compares with that of dendrochronology. But, the δ cycles are obliterated by diffusion in ice older than 10,000 years [Johnsen, 1977].

The distance between two consecutive minima gives the present thickness of the annual layer. When corrected for total strain since the time of deposition, the annual layer thicknesses reveal a record of past accumulation rates. The correction procedure is relatively simple in Central Greenland, and the Milcent and Crête (71.1°N; 37.3°W) records through the last millenium indicate only 5% variance of the accumulation rates, when the records are smoothed with a digital low pass filter of 120 yrs cut off period [Reeh et al., 1978]. In South Greenland the correction procedure is more complicated, and we are not sure that the almost 20% variance found at Dye 3 is real.

The smoothing is necessary also because of the non-uniform areal distribution of snow accumulation, if only due to the sastrugies, which introduce considerable noise. The signal to noise ratio, f_λ, has been calculated at some stations, where two or more cores were drilled. When considering the individual annual layer thickness, the signal to noise ratio is around 4 in South Greenland and only 1 or 2 in Central Greenland (Figure 2) [Reeh and Fisher, 1982]. A similar relationship is valid for mean annual δ values which means that, in order to obtain records that are representative of a large area in Central Greenland, a time series must be smoothed by a low pass filter of longer cut-off period than necessary in South Greenland.

Another question is: to what degree are smoothed δ records representative of climatic changes in terms of secular surface temperature variations? Like any other time series of proxy data, including the deep sea foraminifera records, the δ profiles along ice cores are influenced by many parameters, for example the summer to winter precipitation ratio, and the mean condensation temperature in precipitating clouds. At present, there exists a linear relationship in Greenland between mean δ of the snow pack and mean annual air temperature [Dansgaard et al., 1973]. Climatic interpretation of δ records implies, for one thing, that a similar relationship is valid for temporal temperature changes. Until the general circulation in the atmosphere and in the oceans, and their interactions with climatic changes, are better understood, it is considered premature to try to develop a model that allows extraction of a pure temperature signal from the δ records. Therefore, all we can do at present to check their climatic relevance in terms of temperature is to follow the procedure used on other proxy data: compare them with observed temperature records as far as they reach, which is only a few hundred years, and with other proxy data beyond that limit.

The last 500 yrs of the Crête δ record, smoothed by a 60-year low

[1] Now at: Science Institute, University of Iceland, Reykjavik, Iceland.

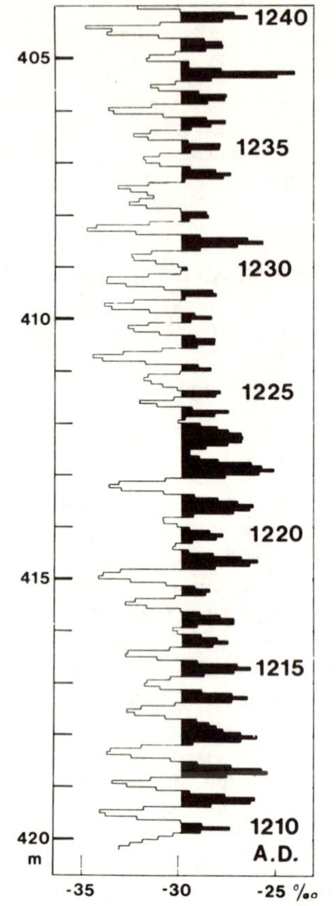

Fig. 1. Seasonal δ variations in the deepest part of the 400 m long ice core from Milcent, mid Greenland. The depth scale has been corrected for vertical strain since the time of deposition. The dating shown to the right was accomplished by counting summer δ peaks downwards from the surface.

pass filter, is plotted to the left in Figure 3. The mid section shows observed Iceland temperatures back to 1860, and systematically observed sea ice occurrences back to 1590, calibrated in terms of mean annual air temperatures [Bergthorsson, 1969]. The right section shows observed Mid England temperatures back to 1690 [Manley, 1961], and estimated 50-year mean air temperatures [Lamb, 1968]. The three curves correlate significantly as long as we keep to systematically observed data. All of them reveal the warm periods in our own century and in the eighteenth century. But, apparently, the so-called little ice age did not culminate in the seventeenth century in mid Greenland, nor in Iceland, where it is often defined as a cold period that lasted from the end of the medieval to the beginning of our century.

The Pleistocene to Holocene δ Shift

The 1400 m long Camp Century (77.2°N; 61.2°W) δ record [Johnsen et al., 1972] was exciting because, first of all, in the deepest 300 m that is plotted on a linear distance-from-bedrock scale to the right in Figure 4, the entire Wisconsin glaciation was revealed in hitherto unseen detail. The Pleistocene to Holocene shift in δ is no less than 11 ‰ compared with 7 ‰ in the new 2037 m long ice core from Dye 3 (cp. the left section of Figure 4).

The δ-shift has been measured on a number of other ice cores that reach back into the last glaciation, and all of the δ-shifts are listed in Table 1. With some exceptions in the northern Baffin Bay area, δ seems to have increased by 5.5 to 7 ‰ at high northern as well as high southern latitudes.

As to the anomalous δ shifts found at several locations in the Baffin Bay area (cp. values above the horizontal line in Table 1), Fisher et al., [1983] recently demonstrated a scouring effect close to the crest of the small Agassiz ice cap, Ellesmere Island, where winter snow particularly is blown away from steep slopes by katabatic winds, or by turbulence caused by a well defined ridge crest, thus influencing the mean δ value of the snow pack. Obviously the scouring effect depends on the shape of the snow surface and may therefore be a contributary reason for anomalous Pleistocene to Holocene δ-shifts in ice cores from relatively small ice sheets. However, it seems unlikely that substantial scouring occurs, or ever

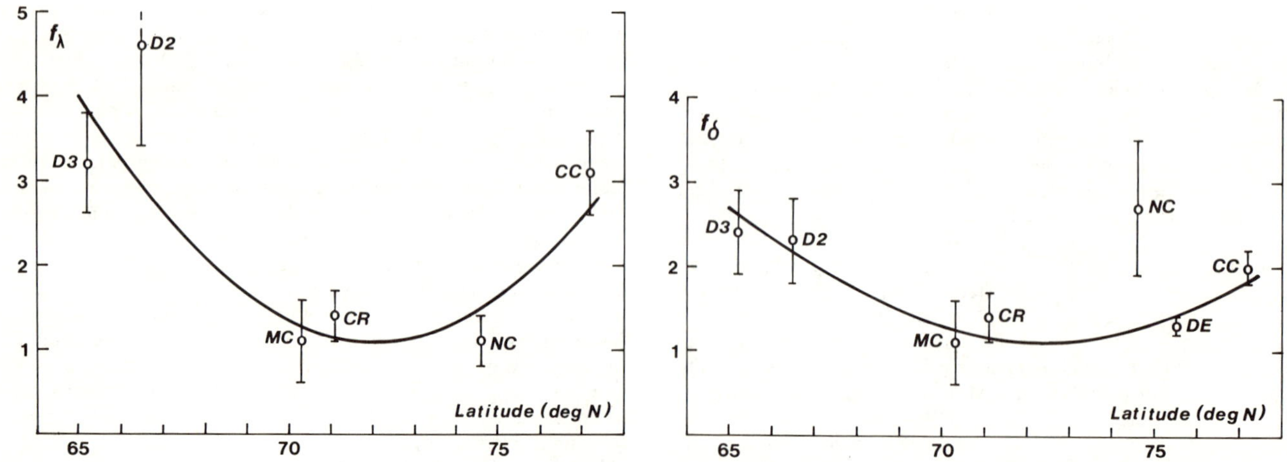

Fig. 2. Signal to noise ratios in time series of annual layer thicknesses (left) and mean annual δ values (right) plotted as a function of latitude. D3 = Dye 3; D2 = Dye 2; MC = Milcent; CR = Crête; NC = North Central; DE = Devon Island; CC = Camp Century. (From Fisher et al., 1985.)

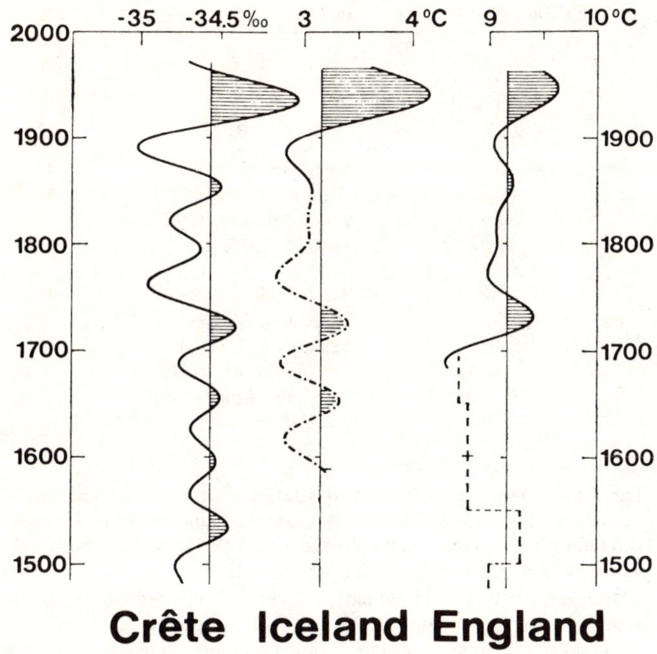

Fig. 3. Correlation between a smoothed and absolute dated δ record from Crête, and mean air temperatures in Iceland and England. Solid curves are based on direct observations; the dashed-dotted curve is based on systematic, indirect observations.

occurred, along the ridges of the Greenland and Antarctic ice sheets, because their surface slopes close to the ridges are an order of magnitude smaller than those of the Agassiz ice cap, and the ridge crests are correspondingly less well defined.

The highest δ-shift (15 ‰) listed in Table 1 is probably due in part to the Barnes ice cap being overrun by the great Laurentide ice sheet during the glaciation. The ice from this period therefore originates from high surface elevations far inland [Hooke and Clausen, 1982]. Even the minor δ-shifts (5–7 ‰) may in part be due to changing elevation of the site of snow deposition during the climatic transition, either because the ice surface elevation generally decreased relative to the (rising) sea level, or because the Holocene part of the ice core was deposited much closer to the drill site, and hence at lower elevations, than older ice. However, model calculations of the present and past ice flow in the Greenland ice sheet [Reeh, 1982; 1984] show that, even if an ice bridge connected the ice sheets in NW Greenland and the Canadian Arctic islands [Dansgaard et al., 1973], the elevation of the snow deposition sites hardly changed by more than a few hundred meters in NW Greenland, and less in case of no such ice bridge (the existence of which has been contradicted by geological evidence [England and Bradley, 1978]). The model calculations also show that the deep ice in the Camp Century core cannot have traveled very far since the time of deposition and, consequently, the high Camp Century shift in δ (Table 1), has to be explained by effects other than substantial shift in the elevation of the site of deposition.

This conclusion seems to be inconsistent with an observed 15 percent increase in total gas content during the climatic transition, which was interpreted as a result of a 1200-m lowering of the surface elevation [Raynaud and Lorius, 1973], or 900 m when corrected for temperature change [Raynaud and Lebel, 1979]. However, the relation between total gas content and elevation is not fully clarified [Dansgaard et al., 1973, p. 34; Paterson, 1981, p. 343]. For example, one has to assume that the present temperature to grain size relationship was valid in the past. This may be true in Antarctica, but not necessarily in the case of Camp Century, where the dust concentration is one to two orders of magnitude lower in the Holocene ice than in the Wisconsin ice [Hammer, 1977; Thompson, 1977]. Since the dust particles serve as condensation nuclei for the

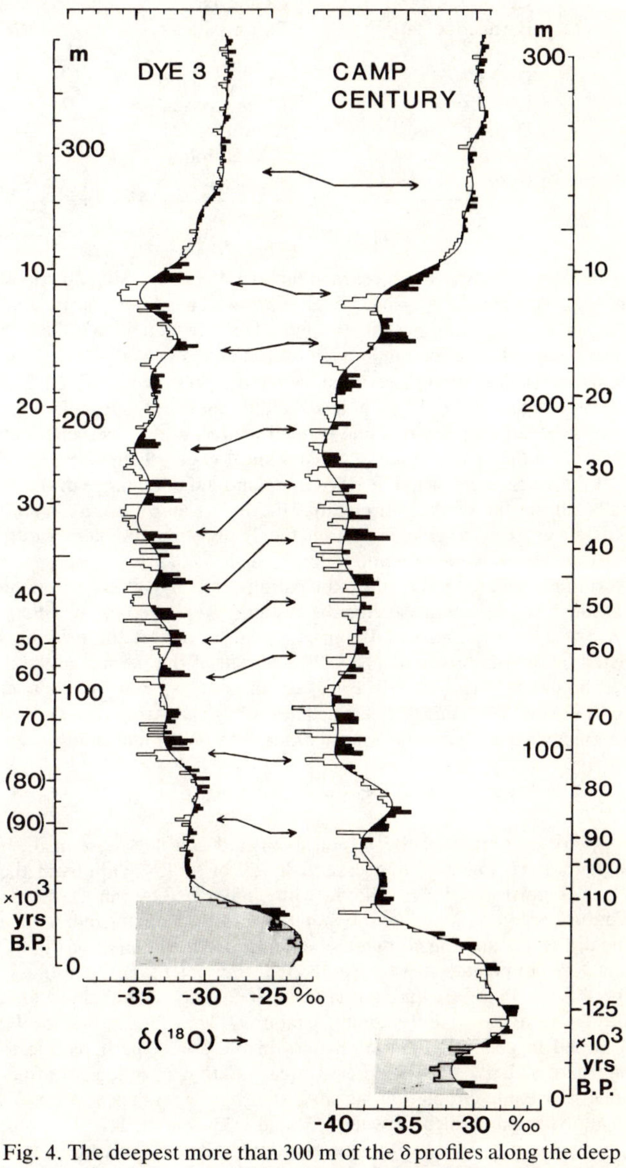

Fig. 4. The deepest more than 300 m of the δ profiles along the deep ice cores from Dye 3 and Camp Century plotted in 1 m increments as functions of the distance from bedrock. The smoothed curves are obtained by digital low pass filters that cut off all frequencies higher than one corresponding to wave lengths of 24 and 20 m, respectively. The shaded areas symbolize silty ice. The arrows show some common features in the two records. The dashed line indicates that the increment from 29 to 30 m above bedrock at Camp Century contains an only 15-cm thick layer with a δ value as low as -29 ‰, corresponding to full glacial severity. The outer time scales are transferred from Figure 5.

TABLE 1. Location of 16 Cores Extending in Time Back to the Wisconson Stage

Drill site	Region	Latitude	Pleistocene to Holocene δ shift (‰)	Ref.
Agassiz ice cap	Ellesmere Isl.	81°N	10	Fisher et al. (1983)
Camp Century	NW Greenland	77°N	11–12	Johnsen et al. (1972)
Devon ice cap	Devon Isl.	75.5°N	7	Paterson et al. (1977)
Barnes ice cap	Baffin Isl.	69.7°N	15	Hooke (1976)
Dye 3	SE Greenland	65.2°N	7	This work
Law Dome	E. Antarctica	66.5°S	7	Budd and Morgan (1977)
Dome C	E. Antarctica	74.6°S	5–6	Lorius et al. (1979)
Vostok	E. Antarctica	78.5°S	5–6	Gordienko et al. (1982)
Byrd	W. Antarctica	80°S	7	Johnsen et al. (1972)

early recrystallization processes in the firn, the decreasing dust load through the climatic transition might very well have been associated with increasing grain size in the firn. The consequent increase in pore volume in the ice could account for part, or maybe all, of the observed shift in total gas content. Furthermore, the scouring effect mentioned above would seem to invalidate the method on minor ice sheets, because removal of fine grained winter snow causes elevated total gas content in the ice [Raynaud and Lebel, 1979].

If the surface elevation in NW Greenland did not change dramatically during the climatic transition, the high Camp Century δ-shift must be ascribed to either an anomalously high climatic temperature shift in this area, or to a anomalously high δ change per degree of warming. Whatever the full explanation might be, there are good reasons to believe that the climatic conditions changed more radically in the northern part of Baffin Bay than at any of the drill sites listed below the line in Table 1. The retreat of the permanent pack ice boundary brought the open ocean at least 2000 km closer [CLIMAP, 1981], and the deglaciation of North America removed an efficient barrier against warm winds from West and South.

The Deepest 300 m of the Records

As to the details of the δ records in mid and late Wisconsin, it appears from Figure 4 that essentially all of the oscillations in the Dye 3 record may be correlated with similar features in the Camp Century record. In view of the more than 1400 km distance between the drill sites, and the completely different ice flow conditions in SE and NW Greenland, it is concluded that the violent δ oscillations in the Wisconsin ice are hardly due to any local climatic effect in Baffin Bay, nor to any ice-dynamic instability. They are rather to be ascribed to general climatic changes in the arctic, perhaps due to alternations between two different quasi-stationary modes of atmospheric/oceanic circulation, as suggested by the tendency of the δ's to jump back and forth between −31 and −35 ‰ in the Dye 3 record, and between −38 and −42 ‰ in the Camp Century record.

The correlation can hardly be extended any closer to the bottom than $y = 50$ m at Dye 3, respectively $y = 76$ m at Camp Century, and even this may be too far, because a silty layer of apparent bedrock origin was found at $y = 87$ m in the Dye 3 core. From $y = 50$ m the ice at Dye 3 is almost homogeneous in δ all the way down to the silty ice, whereas the Camp Century ice continues to vary in δ below $y = 76$ m.

Independent dating of the Wisconsin ice in Greenland is a problem. So far, no radio-isotope dating method has been adequately developed to be applied on the limited amounts of ice available in ice cores [Dansgaard, 1981] and stratigraphic methods are hardly applicable beyond 10,000 yrs BP, as mentioned above. Ice flow modelling is inadequate due to the lack of knowledge about past accumulation rates and, in the case of the Dye 3 area, because of an extremely complicated flow pattern over the hilly bedrock upstream (cp. Overgaard and Gundestrup [1984], Reeh et al. [1984], Jezek et al. [1984], and time scale correction procedures based on Fourier spectral analyses [Dansgaard et al., 1971] call for an initial dating close to absolute chronology. Radioactive dating of deep ice cores may be possible in the near future by accelerator based mass spectrometry. However, for the moment it may be worthwhile to aim at indirect dating of the deep parts of the ice cores by comparison with marine records.

Since the early attempts to date the Camp Century ice core [Dansgaard and Johnsen, 1969; Dansgaard et al., 1970; 1971], comprehensive studies of sediment cores from the deep oceans have led to a relatively consistent picture of the climatic development through the last glacial cycle, i.e. back to the Eem/Sangamon interglacial some 125,000 years ago. As an example, the right section of Figure 5 shows the upper 4 m (corresponding to the last 132,000 years) of one of the most intensively studied deep sea records: Hays, Imbrie and Shackleton's [1976] ^{18}O analyses of planktonic foraminifera in a core from sub-polar Indian Ocean, some 18,000 km from Camp Century. Increasing δ's, toward left, correspond to increasing continental ice volume. The Emiliani [1966] stages shown along the record refer to distinct glaciation features that are recognizable in most deep sea records.

The smoothed version of the Camp Century δ profile from Figure 4 is shown in the left section of Figure 5. The tentative correlation indicated by the arrows suggests a close relationship between the two records. Since there is hardly more than a few thousand years time lag between changes in climate and ice volume, the marine time scale has been transferred to the Camp Century record in accordance with the arrows, except for the 10,000-year mark that is taken from our own measurements of annual layers [Hammer et al., 1978]. In turn, the new tentative time scale is transferred to the detailed records in Figure 4. Obviously, the new Camp Century time scale is not linear, by far. The 10,000-year depth intervals do not even decrease monotonically downwards, as they would do in case of a very simple ice flow and accumulation pattern in the past.

If we assume that varying accumulation rate is the main reason for these irregularities, they would unambiguously indicate relatively high accumulation rates in the 125–115,000, the 80–60,000, and the 40–30,000-year intervals. These are the intervals, during which the deep sea records (cp. Figure 5) show that considerable build-up of

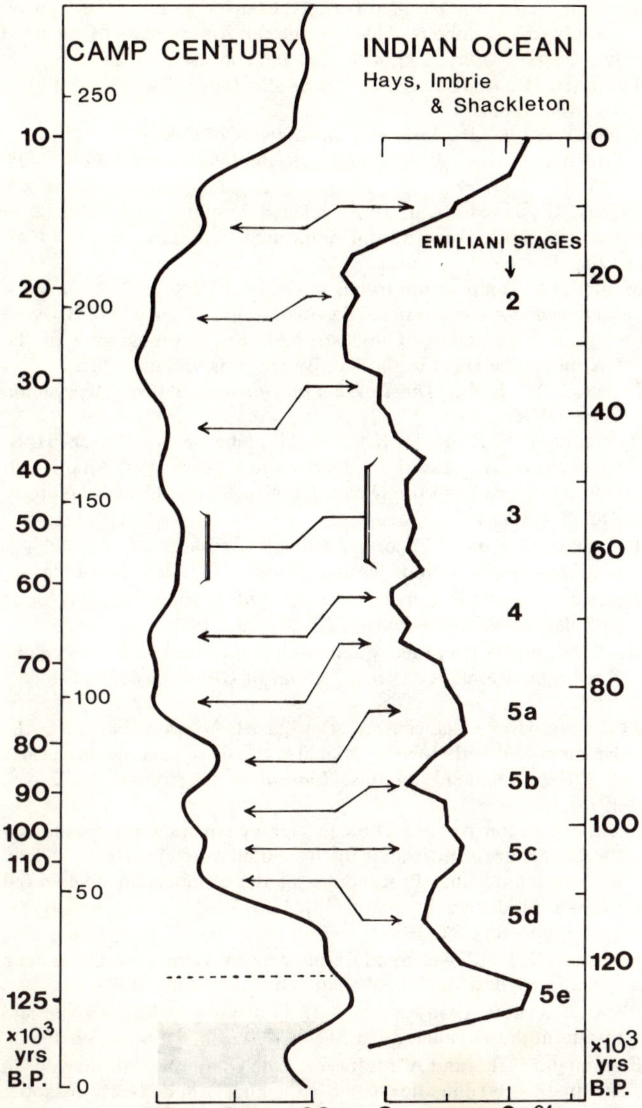

Fig. 5. The smoothed version of the deepest 270 m of the Camp Century δ profile, compared with the upper 4 m of a planktonic foraminifera δ profile from South Indian Ocean. The linear time scale along the deep sea core is transferred to the Camp Century ice core in accordance with the correlation suggested by the arrows, but for the 10,000 yr BP mark to the upper left, which originates from measurements of annual layer thicknesses [Hammer et al., 1978]. (From Dansgaard et al., [1982].)

continental ice occurred. It is particularly interesting that the apparently high accumulation rates after the dramatic cold spell at 120,000 BP were associated with δ's only slightly lower than Eem/Sangamon values. This agrees with the presence of warm planktonic foraminifera in marine North Atlantic sediments deposited in the same period of ice growth [Ruddiman and McIntyre, 1979], which suggests that the sub-polar North Atlantic Ocean maintained warm surface temperatures long after the ice began to pile up on the continents in the northern hemisphere. All of this—an extremely cold spell followed by a long period of high accumulation—would seem to illustrate how it was possible, within the 5–10,000 years that succeeded the Eem/Sangamon, to build up some 25,000,000 cu km of continental ice, corresponding to a 70-m lowering of the sea level, as indicated by the deep sea record.

However, the ice flow history of the ice cap in NW Greenland is a potential contributary reason for the irregular Camp Century time scale, and this contribution can be estimated only by more realistic ice flow modelling than is possible today. The accumulation rate considerations should therefore be taken with some reservation, and they are presented here mainly to demonstrate the kind of valuable information that could be obtained from a δ profile along a deep ice core, drilled on a scientifically favorable location, i.e. in an area of:

1. *simple ice flow history* that will facilitate realistic modelling;
2. *high accumulation rate*, at present more than 25 cm ice/yr, to ensure continuity of the time scale, also under glacial conditions, and preservation of annual δ cycles at least back to 10,000 yrs BP, and
3. *bedrock temperatures below the pressure freezing point*, which ensures a long time range.

Central Greenland meets these fundamental requirements. An ice core to bedrock in this area would reach several hundred thousand, perhaps a million years back in time, and several glacial cycles might be studied in fine detail by isotope and impurity analyses.

Acknowledgments. We are grateful for the financial, logistic and other support provided by U.S. National Science Foundation, Division of Polar Programs, Commission for Scientific Research in Greenland, Danish Natural Science Research Council, Carlsberg Foundation, and Nordic Cultural Foundation. Furthermore, we thank Polar Ice Coring Office, University of Nebraska for logistic management, and the numerous individuals, who helped us in the field and in the laboratories and workshops involved.

References

Barkov, N. J., F. G. Gordienko, E. S. Korotkevich and V. M. Kotlyakov, Oxygen isotope studies on a 500 m ice core from a bore hole at Vostok station, *Information Bulletin Soviet Antarctic Expedition*, *90*, 39–48, 1975.

Bergthorsson, P., Sea ice in Iceland waters, *Jökull* (Reykjavik), 19, 94–115, 1969.

Budd, W. F., and V. J. Morgan, Isotopes, climate and ice sheet dynamics from core studies on Law Dome, Antarctica, Symposium on Isotopes and Impurities in Snow and Ice, (Proceedings of the Grenoble Symposium, 1975), *IAHS-AISH Publication No. 118*, 312–321, 1977.

CLIMAP project members, Seasonal reconstructions of the earth's surface at the last glacial maximum, *Geological Society of America, Map and Chart Series*, *36*, 1981.

Dansgaard, W., Dating the past to find the future, *Nature*, *290*, 360–361, 1981.

Dansgaard, W. and S. J. Johnsen, A flow model and a time scale for the ice core from Camp Century, Greenland, *Journal of Glaciology*, *8*, 215–223, 1969.

Dansgaard, W., S. J. Johnsen, H. B. Clausen and C. C. Langway, Jr., Ice cores and palaeoclimatology, In *Radiocarbon Variations and Absolute Chronology*, edited by I. U. Olsson, XII Nobel Symposium, Almquist & Wiksell, Stockholm, pp. 337–351, 1970.

Dansgaard, W., S. J. Johnsen, H. B. Clausen and C. C. Langway, Jr., Climatic record revealed by the Camp Century ice core, in *The Late Cenozoic Glacial Ages*, edited by K. K. Turekian, pp. 37–56, Yale University Press, 1971.

Dansgaard, W., S. J. Johnsen, H. B. Clausen and N. Gundestrup,

Stable isotope glaciology, *Meddelelser om Grønland, 197*, No. 2, 53 pp., 1973.

Dansgaard, W., H. B. Clausen, N. Gundestrup, C. U. Hammer, S. J. Johnsen, P. M. Kristinsdottir and N. Reeh, A new Greenland deep ice core, *Science, 218*, 1273–1277, 1982.

Emiliani, C., Paleotemperature analysis of Caribbean cores P 6304-9 and a generalized temperature curve for the past 425,000 years, *Journal of Geology, 74*, 109–174, 1966.

England, J., and R. S. Bradley, Past glacial activity in the Canadian High Arctic, *Science, 200*, 265–270, 1978.

Epstein, S., and C. Benson, Oxygen isotope studies, EOS *Transactions, American Geophysical Union, 40*, 81–84, 1959.

Fisher, D. A., R. M. Koerner, W. S. B. Paterson, W. Dansgaard, N. Gundestrup and N. Reeh, Effect of wind scouring on climatic records from ice-core oxygen-isotope profiles, *Nature, 301*, 205–209, 1983.

Fisher, D. A., N. Reeh and H. B. Clausen, Stratigraphic noise in time series derived from ice cores. Annals of Glaciology, in press, 1985.

Gordienko, F. G., V. M. Kotlyakov, N. I. Barkov and S. Korotkevich, Results of oxygen-isotope studies on the Vostok core. Paper presented in the IAGP Scientific Session, House of Scientists, Leningrad, 6 July, 1982.

Hammer, C. U., Dust studies on Greenland ice cores, Symposium on Isotopes and Impurities in Snow and Ice, (Proceedings of the Grenoble Symposium, 1975), *IAHS-AISH Publication No. 118*, 365–370, 1977.

Hammer, C. U., Acidity of polar ice cores in relation to absolute dating, past volcanism, and radio-echoes, *Journal of Glaciology, 25*, 359–372, 1980.

Hammer, C. U., H. B. Clausen, W. Dansgaard, N. Gundestrup, S. J. Johnsen and N. Reeh, Dating of Greenland ice cores by flow models, isotopes, volcanic debris, and continental dust, *Journal of Glaciology, 20*, 3–26, 1978.

Hammer, C. U., H. B. Clausen and W. Dansgaard, Greenland ice sheet evidence of post-glacial volcanism and its climatic impact, *Nature, 288*, 230–235, 1980.

Hays, J. D., J. Imbrie and N. J. Shackleton, Variations in the Earth orbit: pacemaker of the ice ages, *Science, 194*, 1121–1132, 1976.

Hooke, R. LeB., Pleistocene ice at the base of the Barnes Ice Cap, Baffin Island, N.W.T., Canada, *Journal of Glaciology, 17*, 49–60, 1976.

Hooke, R. LeB., and H. B. Clausen, Wisconsin and Holocene $\delta^{18}O$ variations, Barnes ice cap, Canada, *Geological Society of America Bulletin, 93*, 784–789, 1982.

Jezek, K. C., E. A. Roeloffs and L. L. Greischar, A geophysical survey of subglacial geology around the deep-drilling site at Dye 3, Greenland, this volume, 1985.

Johnsen, S. J., Stable isotope homogenization of polar firn and ice, Symposium on Isotopes and Impurities in Snow and Ice, (Proceedings of the Grenoble Symposium, 1975), *IAHS-AISH Publication No. 118*, 210–219, 1977.

Johnsen, S. J., W. Dansgaard, H. B. Clausen and C. C. Langway, Jr., Oxygen isotope profiles through the Antarctic and Greenland ice sheets, *Nature, 235*, 429–434, and *236*, 249, 1972.

Lamb, H. H., *The Changing Climate*, Section 7, 236 pp., Methuen, London, 1968.

Langway, C. C. Jr., Stratigraphic analysis of a deep ice core from Greenland, *Geological Society of America, Special Paper 125*, 186 pp., 1970.

Lorius, C., L. Merlivat, J. Jouzel and M. Pouchet, A 30,000 yr isotope climatic record from Antarctic ice, *Nature, 280*, 644–648, 1979.

Manley, G., Temperature trends in England 1698–1957, *Archive for Meteorology, Geophysics, Bioclimatology, B9*, p. 413–433, 1961.

Overgaard, S., and N. Gundestrup, Bedrock topography of the Greenland ice sheet in the Dye 3 area, this volume, 1985.

Paterson, W. S. B., *The Physics of Glaciers*, 380 pp., Pergamon Press, 1981.

Paterson, W. S. B., R. M. Koerner, D. Fisher, S. J. Johnsen, H. B. Clausen, W. Dansgaard, P. Bucher and H. Oeschger, An oxygen-isotope record from the Devon Ice Cap, Arctic Canada, *Nature, 266*, 508–511, 1977.

Raynaud, D., and C. Lorius, Climatic implication of total gas content in ice at Camp Century, *Nature, 243*, 283–284, 1973.

Raynaud, D., and B. Lebel, Total gas content and surface elevation of polar ice sheets, *Nature, 281*, 289–291, 1979.

Reeh, N., A plasticity theory approach to the steady-state shape of a three-dimensional ice sheet, *Journal of Glaciology*, 28, 431–455, 1982.

Reeh, N., H. B. Clausen, W. Dansgaard, N. Gundestrup, C. U. Hammer and S. J. Johnsen, Secular trends of accumulation rates at three Greenland stations, *Journal of Glaciology, 20*, 27–30, 1978.

Reeh, N., Reconstruction of the glacier ice covers of Greenland and the Canadian Arctic islands by three-dimensional perfectly plastic ice sheet modeling (Proceedings of the Symposium on Ice and Climate Modeling, Evanston, Ill., U.S.A., 27 June–1 July, 1983, vol. 5, pp. 115–121, 1984.

Reeh, N., S. J. Johnsen and D. Dahl-Jensen, Dating the Dye 3 deep ice core by flow model calculations, this volume, 1985.

Risbo, T., H. B. Clausen and K. L. Rasmussen, Supernovae and nitrate in the Greenland Ice Sheet, *Nature, 294*, 637–639, 1981.

Ruddiman, W. F., and A. McIntyre, Late Quaternary surface ocean kinematics and climatic change in the high latitude North Atlantic, *Journal of Geophysical Research, 82*, 3877–3887, 1977.

Thompson, L. G., Variations in microparticle concentration, size distribution and elemental composition found in Camp Century, Greenland, and Byrd Station, Antarctica, ice cores, Symposium on Isotopes and Impurities in Snow and Ice, (Proceedings of the Grenoble Symposium, 1975), *IAHS-AISH Publication No. 118*, 351–364, 1977.

CHLORIDE, NITRATE, AND SULFATE IN THE DYE 3 AND CAMP CENTURY, GREENLAND ICE CORES

Michael M. Herron[1] and Chester C. Langway, Jr.

State University of New York at Buffalo, Department of Geological Sciences, Ice Core Laboratory, Amherst, New York

Abstract. Concentrations of Cl^-, NO_3^-, and SO_4^{2-} were measured over the entire depth interval of the deep ice cores from Dye 3 and Camp Century. Using a preliminary time scale for the Dye 3 core and the published Camp Century time scale, the chemical profiles reveal dramatic, synchronous changes in precipitation composition over the last 90,000 years. The Holocene-Wisconsin boundary is marked by a sudden increase in concentrations with increasing depth of all constituents. In the Dye 3 core this transition occurs over an interval of less than 50 cm for SO_4^{2-} and NO_3^-, and less than 200 cm for Cl^- at a depth of 1786 m. Late Wisconsin ice is characterized by lower NO_3^- and higher Cl^- and SO_4^{2-} concentrations compared to Holocene levels. Long periods of intensive volcanic activity are indicated by sporadic SO_4^{2-} and lesser Cl^- peaks from 10,000 to 20,000 BP and 25,000 to 30,000 BP. The background SO_4^{2-} concentrations in the Dye 3 core indicate that Late Wisconsin snow accumulation rates were as little as 22% of the present rate, assuming a constant SO_4^{2-} flux. A method is outlined whereby the paleoelevations and perhaps the rate of disintegration of unstable ice masses, such as the West Antarctic Ice Sheet, can be obtained from Cl^- concentration profiles.

Introduction

It has long been recognized that polar ice sheets contain potentially valuable information on past climates and environments in the form of the chemical content of the snow and ice strata [Murozumi et al., 1969; Langway, 1970]. The precipitation falling on polar ice sheets is the purest in the world due to the ice sheets' elevation and remoteness from atmospheric impurity sources. In spite of low impurity concentrations, clean sample handling techniques and accurate analytical procedures have been developed that are beginning to unlock the secrets contained in the chemical composition of past snowfalls.

Since the application of ion chromatography to glaciochemical analysis [Herron, 1980], a great deal of information has evolved on the Cl^-, NO_3^-, and SO_4^{2-} content of recent Greenland and Antarctic precipitation [M. Herron, 1982]. The understanding of the impurity sources and the parameters that influence the anion concentrations in recent polar snow forms a framework for the interpretation of the concentration profiles of the deep ice cores from Dye 3 and Camp Century, Greenland. The framework may be summarized as follows:

Chloride—The overwhelmingly dominant source of Cl^- in polar precipitation is sea-salt aerosols, although some large volcanic eruptions produce detectable Cl^- concentration perturbations. Holocene Cl^- concentrations are primarily dependent on site elevation, with concentrations decreasing logarithmically as elevation increases. For sites of equal elevation, Antarctic Cl^- concentrations are slightly (20–50%) greater than Greenland concentrations.

Sulfate—A significant anthropogenic SO_4^{2-} component is detectable in recent Greenland snow, but not in Antarctic snow. For ice deposited prior to the industrial impact, there is a relatively constant background SO_4^{2-} component of presumed biogenic origin. Following major eruptions, SO_4^{2-} concentrations rise above this background level for periods up to several years. Background SO_4^{2-} concentrations are inversely proportioned to the snow accumulation rate, A. That is, the background SO_4^{2-} flux is constant over large regions with the Greenland flux about double that in Antarctica.

Nitrate—A significant anthropogenic NO_3^- component is also detectable in recent Greenland snow, but not in the Antarctic. The source of background NO_3^- in pre-industrial ice is still unknown. The source is apparently more active in the northern hemisphere since, for comparable sites, Greenland concentrations are about double those in Antarctica. In both hemispheres there is a distinct summer concentration maximum that renders NO_3^- suitable for dating. Finally, background NO_3^- concentrations appear to decrease with increasing accumulation rate, though not with the inverse relationship exhibited by SO_4^{2-}.

With these empirical relationships, the long-term changes in Cl^-, SO_4^{2-}, and NO_3^- concentrations can be examined in the deep ice cores.

Analytical Procedures

The Dye 3 core sampling and analysis procedures have been detailed elsewhere [M. Herron, 1982]. Intervals of approximately ¼ core were cut by band saw. Surface contamination, including residual drilling fluid, was removed by alternate rinsings of acetone and deionized, filtered water. The cleaned samples were placed in precleaned polystyrene cups with tight-fitting lids to minimized atmospheric modification, melted in a microwave oven, and analyzed for anion concentrations by ion chromatography. Most of the Dye 3 analyses were performed on site in a heated laboratory equipped with a water system and Class 100 clean air station. The remainder of the Dye 3 frozen samples and all of the Camp Century frozen samples were analyzed at the SUNY/Buffalo facility which includes a Class 100 clean room. Analytical precision is 20% for the Camp Century samples and 10% for the Dye 3 samples at the 2σ level of confidence.

The Dye 3 samples were from 155 depth intervals with as many as 70 samples per depth interval in the shallower samples and generally

[1] Now at: Schlumberger-Doll Research, Ridgefield, Connecticut.

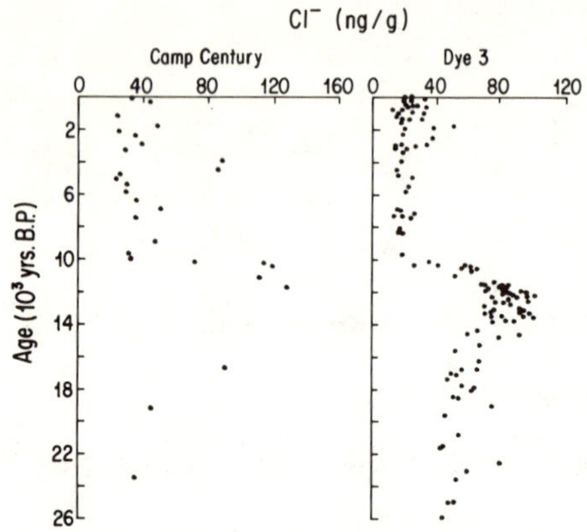

Fig. 1. Cl^- concentrations as a function of estimated age in the 0–26 kBP sections of the Camp Century and Dye 3 ice cores.

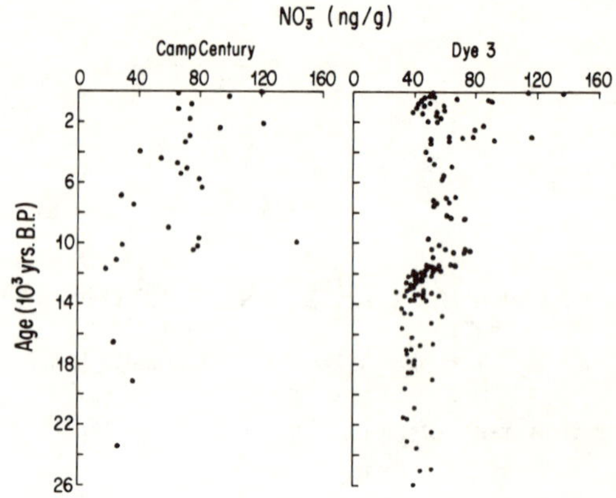

Fig. 3. NO_3^- concentrations as a function of estimated age in the 0–26 kBP sections of the Camp Century and Dye 3 ice cores.

a single sample in deeper ice where annual layers are very thin. The Camp Century analyses are more in the nature of a preliminary survey, with 38 depth intervals sampled and two to nine samples taken at each interval. Ages for the Camp Century core are calculated from the model of Hammer et al. [1978]. Ages and annual layer thicknesses for the Dye 3 core are calculated from the preliminary time scale of Dansgaard and Reeh [1982].

Results

Profiles 0-26 kBP

Figures 1, 2, and 3 show the Cl^-, SO_4^{2-}, and NO_3^- concentration profiles for the Dye 3 and Camp Century ice cores to an age of 26 kBP. At a depth of about 1950 m in the Dye 3 cores, with calculated age of 26 kBP, debris of probable subglacial origin were encountered. Inasmuch as stratigraphic continuity below this depth cannot be guaranteed, the profiles will first be given for ice above this depth.

The Cl^- concentration profiles are shown in Figure 1. The detailed Dye 3 results show a fairly steady Holocene baseline concentration of about 18–20 ng/g, punctuated occasionally by volcanogenic Cl^- contributions. The Camp Century results show more scatter with a Holocene average of about 34 ng/g. Two

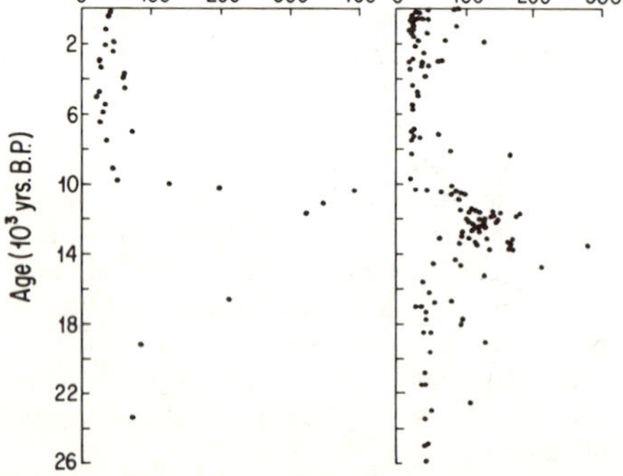

Fig. 2. SO_4^{2-} concentrations as a function of estimated age in the 0–26 kBP sections of the Camp Century and Dye 3 ice cores.

pronounced peaks at about 4 kBP reflect prehistoric eruptions which are also evident in the SO_4^{2-} profile. The relative baseline Cl^- concentrations are consistent with the differences in site elevation with the higher site, Dye 3, receiving less sea-salt aerosol. The Cl^- profiles indicate a fairly stable concentration at both sites for the past 10,000 years.

Ice from the very end of the Wisconsin stage contains significantly greater Cl^- concentrations than does Holocene ice. For Dye 3 ice of calculated ages 11.0–14.0 kBP, Cl^- concentrations are consistently 3.7–5.3 times greater than the background Holocene level. There are far fewer points for comparison in the Camp Century profile, but Late Wisconsin ice has about 3.4 times as much Cl^- as does Holocene ice. Chloride concentrations in the Dye 3 core are consistently greater than Holocene values for all samples older than 10.4 kBP.

The transition from high Cl^- concentrations to Holocene levels occurs suddenly in the Dye 3 core at a calculated age of 10.4 kBP. This age is about 400 years older than the age of the transition at Camp Century. The time scales are imprecise enough that the transition may have occurred simultaneously at both sites. High sea-salt concentrations also characterize Late Wisconsin ice at Byrd Station in West Antarctica [Cragin et al., 1977] and at Dome C in East Antarctica [Petit et al., 1981]. The transition at Byrd is estimated at between 12.0 and 13.0 kBP, while at Dome C the transition age is estimated at approximately 10–12 kBP. We believe that the transition represents an isochron that can be used for cross-correlation between all four cores.

The SO_4^{2-} concentration profiles over the last 26,000 years are shown in Figure 2. Both profiles show elevated concentrations in modern snow; these are primarily attributed to coal combustion [M. Herron, 1982]. Aside from the anthropogenic impact, the profiles both show the fairly constant background SO_4^{2-} concentrations for the past 10,000 years and the perturbations of past volcanic eruptions. The Holocene/Late Wisconsin SO_4^{2-} transition is coincident with the Cl^- transition in both cores. The Late Wisconsin/Holocene ratios for SO_4^{2-} are about 6.4 at Dye 3 and, for the more limited Camp Century data, about 7.5. During the Holocene, the background SO_4^{2-} concentrations average about 22 ng/g at Dye 3 and 34 ng/g at Camp Century. These averages are in accord with the suggestion of a constant SO_4^{2-} flux in Greenland, as the snow accumulation rates are $A=49$ cm H_2O/yr at Dye 3 and $A=32$ cm H_2O/yr at Camp Century.

Nitrate concentrations in modern Dye 3 and Camp Century snow (Figure 3) reveal the impact of fossil fuel combustion on Greenland snow composition. Holocene NO_3^- concentrations show considerable scatter, particularly at Camp Century. It is surprising to see, however, that except at the transition, Late Wisconsin NO_3^- concentrations are lower than during the Holocene. Nitrate is the only chemical species yet measured on any deep polar ice core to have lower concentrations during the ice age. There is an increasing trend, however, from 12 kBP until 10.4 kBP in the Dye 3 core.

The source(s) of NO_3^- are not yet known, but it is interesting to note that there is an increasing trend in NO_3^- concentration with time in the Holocene at Camp Century, while there appears to be a decreasing trend or no trend at Dye 3. In addition, the Camp Century samples show what appears to be a cycle with a period of about 4,000 years during the Holocene. No such trend is evident in the Dye 3 core and the apparent cyclicity at Camp Century may not be real.

In general, chemical profiles for the past 26,000 years at Camp Century and Dye 3 show that Late Wisconsin ice is chemically distinct from Holocene ice. Late Wisconsin ice in both cores is characterized by high Cl^- and SO_4^{2-} concentrations and, except very near the transition, low NO_3^- concentrations.

The 1786 m Transition

The climatic transition in the Dye 3 ice core was first identified to occur at 1786 m by space charge measurements conducted in the field (Hammer, personal communication). In order to examine the glaciochemical transition in detail, the anion composition was analyzed continuously over the depth interval between 1784.47 and 1791.5 m; sample lengths over this zone were typically 9 cm. According to the preliminary depth-age model of Dansgaard and Reeh [1982], the thickness of an annual layer is approximately 2.9 cm, so each sample represents about a three-year average, a comparable time span to that of many Holocene data points in Figures 1–3. The detailed anion concentrations profiles through the climatic transition are presented in Figure 4.

Above the transition, chloride concentrations are about 25 ng/g which is just slightly higher than the average throughout the Holocene. However, between 1785 and 1787 m, over a distance representing a time span of only 70 years, the concentrations rapidly increase to values near 60 ng/g and then they remain constant for several meters. In deeper ice the chloride concentrations increase to the Late Wisconsin values of 70–100 ng/g.

The chemical transition is even more dramatic in the sulfate and nitrate concentration profiles. The sulfate values leap from a background level of 25 ng/g to a background level of 60–70 ng/g upon which are superimposed numerous peaks reflecting tremendous volcanism. This transition occurs abruptly at a depth of 1786.2 m and is complete over an interval of less than 50 cm. Similarly, over a depth interval of about 40 cm, nitrate concentrations increase from their characteristic Holocene levels to a concentration of about 78 ng/g which remains fairly stable between 1786.5 and 1789 m. Below 1789 m, the nitrate concentrations gradually decrease with depth.

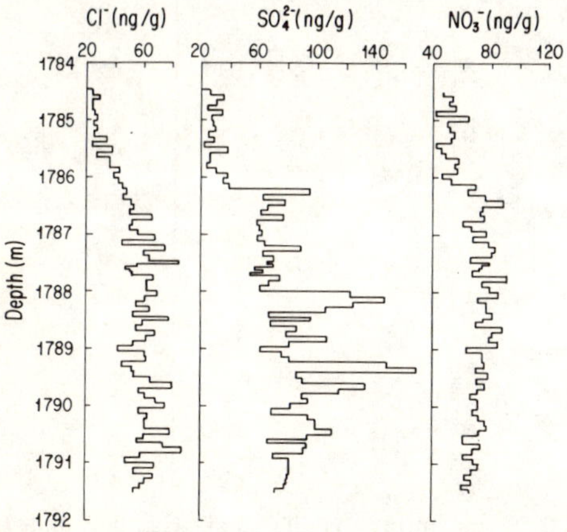

Fig. 4. Cl^-, SO_4^{2-}, and NO_3^- concentrations versus depth at Dye 3 at the 1786 m transition.

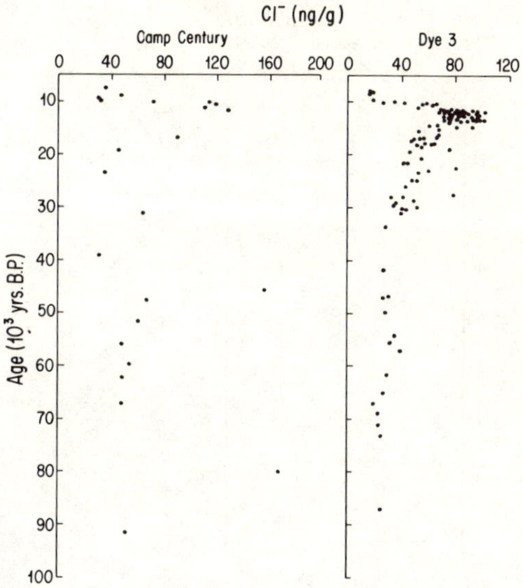

Fig. 5. Cl^- concentrations at Camp Century and Dye 3 for estimated ages >7.5 kBP.

The nearly instantaneous transition from ice that is chemically characteristic of the Late Wisconsin to ice characteristic of the Holocene is one of the most dramatic features of the Dye 3 glaciochemical results. Based on the calculated annual layer thicknesses, most of the chemical transition occurred within not more than 70 years, and perhaps as little as 10 years. Alternate annual layer estimates will increase these time spans by as much as a factor of 2.5, but they still represent surprisingly brief periods of time. Detailed analyses have not been conducted on the other deep ice cores, but the Camp Century results indicate the anionic transition occurred within a five-meter depth interval, also representing about 70 years and thus supporting the Dye 3 results.

Profiles 7.5-92 kBP

Chemical analyses were also carried out on Dye 3 ice below 1950 m, where the first subglacial debris were encountered. Dansgaard and Reeh [1982] indicate that the ice below 1975 m is homogeneous with respect to oxygen isotope, dust, and acidity measurements. The chemical analyses indicate that the ice is not homogeneous.

Figure 5 shows the Cl^- concentration profiles at Dye 3 and Camp Century. In the Dye 3 profile there are Cl^- peaks below the debris layer at calculated ages of 27.5 kBP and approximately 30 kBP. Deeper in the core is a transition with Cl^- values increasing from about 22 ng/g up to about 30 ng/g at a calculated age of about 65 kBP. This transition is not apparent in the Camp Century data, but demonstrates a lack of homogeneity in the deeper ice.

Figure 6 shows the NO_3^- concentration profile at Dye 3 and Camp Century. The Dye 3 results show no discernible break in the pattern at 26 kBP. Instead, pre-55 kBP concentrations are about 50 ng/g, consistent with the Holocene average. For ice with a calculated age

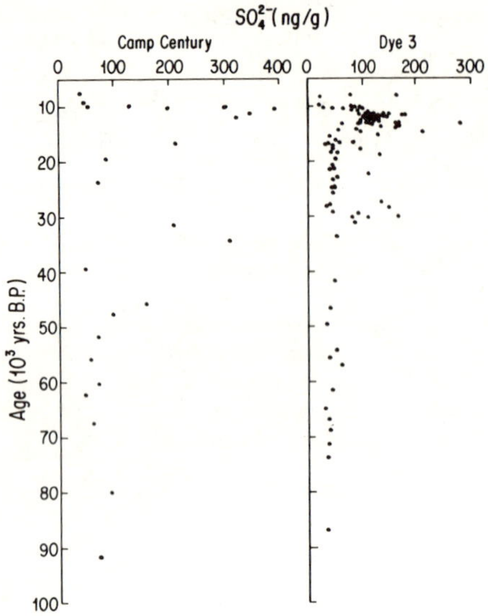

Fig. 6. SO_4^{2-} concentrations at Camp Century and Dye 3 for estimated ages >7.5 kBP.

younger than 55 kBP, Dye 3 NO_3^- concentrations steadily decrease to values less than 40 ng/g in the Late Wisconsin before rising after 12 kBP. To some extent this pattern is followed in deep Camp Century ice.

The SO_4^{2-} profiles in Figure 7 also show that the deep Dye 3 ice is not of uniform composition. Volcanic peaks are indicated at 27.5 kBP and around 30 kBP, in accord with the Cl^- record. Additionally, there is a deep SO_4^{2-} transition in the Dye 3 core with SO_4^{2-} concentrations increasing from about 35 ng/g to about 50 ng/g at a calculated age of about 65 kBP.

Despite the existence of subglacial debris 87 m above the sub-ice boundary, the chemical analyses indicate that underlying ice is not homogeneous and may contain significant paleoenvironmental information. Utilization of this information requires detailed dating using cosmogenic nuclide activities and sophisticated upstream modeling. Because of this, the discussion of the environmental and galciological significance of the Dye 3 chemical data will focus on the ice younger than 26 kBP.

Discussion

Chloride and Elevations

Sea-salt concentrations in the Holocene sections of the Dye 3 and Camp Century ice cores are fairly constant indicating a stable climatological regime. Examination of recent Cl^- concentrations at numerous sites in Greenland and Antarctica suggests that site elevation is the dominant parameter governing seal-salt concentrations [M. Herron, 1982]. Chloride concentrations generally obey the relationship:

$$C = C_0 \exp\left(-\frac{Z}{1.52 \text{ km}}\right) \qquad (1)$$

where C is the measured sea-salt concentration, Z is the site elevation, and 1.52 km is the empirically derived scale height. C_0 is a constant with slightly greater values for Antarctic precipitation compared to Greenland snow which may reflect the activity of sea-salt aerosol generation and the efficiency of transport to the polar ice sheets. Chloride concentrations in the Holocene sections of the Dye 3 and Camp Century ice cores are quite constant and indicate a stable glaciochemical regime for the past 10,000 years.

Chloride concentrations in Late Wisconsin ice are 3.7–5.3 times as high as concentrations in Holocene ice at Dye 3, and 3.4 times as high as in Holocene ice at Camp century. Elevated sea-salt concentrations according to equation (1) imply a greater value of C_0 or lower elevations during the ice age. Reduced site elevations may be dismissed on glaciological grounds. Therefore, the increased sea-salt concentrations most likely reflect increased values of C_0, i.e. higher sea-salt aerosol generation rates and/or increased efficiency of aerosol transport from the oceanic source regions to the polar ice sheets.

A similar explanation was invoked to explain elevated sea-salt concentrations in Late Wisconsin ice from Dome C, Antarctica [Petit et al., 1981]. The concentration ratio of Late Wisconsin ice to Holocene ice was 3.8 for Cl and 4.8 for marine-derived Na. These ratios are essentially equivalent to the ratios observed at Dye 3 and suggest that the magnitude of the C_0 increase may have been a global phenomenon. Using the present relationship between wind speed over the ocean and sea-salt aerosol concentrations [Petit et al., 1981], the Dye 3 and Dome C results both suggest that wind speeds over the oceanic source regions were up to 5–8 m sec^{-1} greater than at present.

The rapid transition in Cl^- concentrations at 1786 m in the Dye 3 core can also be interpreted in terms of wind speeds over oceanic source regions. Chloride concentrations decrease over the transition from about 60 ng/g to about 25 ng/g. This change translates into an irreversible decline in wind speeds of 3.5–5.5 m sec^{-1} within a few tens of years. The suddenness of the transition is echoed in sea-sediment cores which show that the North Atlantic switched from a glacial mode to a largely interglacial mode in less than 2,000 years [Ruddiman and MacIntyre, 1977]. However, an exact estimate of the time span involved cannot be obtained from the sea-sediment cores due to vertical mixing of the sediment. The Dye 3 results indicate that the transition is substantially shorter than 2,000 years.

The observation that present Cl^- concentrations in Greenland and Antarctica are primarily dependent on site elevation suggests that valuable information on past ice-sheet dimensions may be available

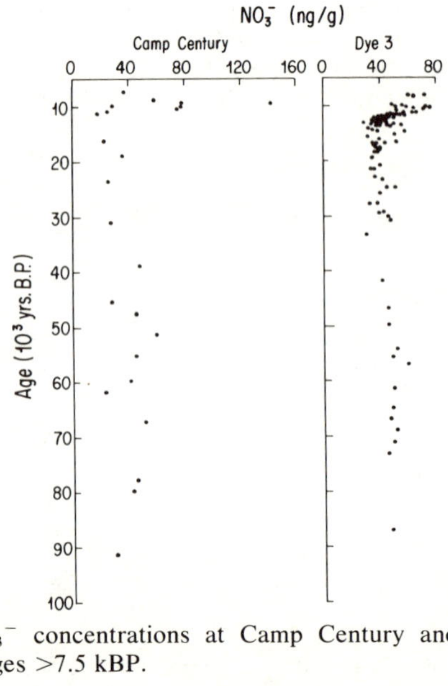

Fig. 7. NO_3^- concentrations at Camp Century and Dye 3 for estimated ages >7.5 kBP.

TABLE 1. Derived Accumulation Rates from Measured SO_4^{2-}

Preliminary Age (10^3 years BP)	Background SO_4^{2-} (ng/g)	Accumulation Rate (kg $m^{-2}a^{-1}$)
0–10	20	490
10.4[a]	25	390
10.4[b]	52	190
11–13	90	110
15–20	42	230
20–55	38	260
60–90	32	310

[a] After 1786 m transition.
[b] Prior to 1786 m transition.

from Cl^- analyses of Late Wisconsin ice. There are at present insufficient data from the other deep ice cores to test this hypothesis, but the potential application will be briefly outlined. It is assumed that site elevation and sea-salt aerosol generation activity are the primary parameters governing Late Wisconsin and Holocene sea-salt concentrations. Following equation (1), sea-salt concentrations as a function of time in a given ice core may be expressed as

$$C(t) = S(t) \exp\left(-\frac{Z(t)}{1.52 \text{ km}}\right) \quad (2)$$

where $S(t)$ describes the source activity and $Z(t)$ describes the changing elevation of origin of ice in an ice core. For sites near an ice dome or ice crest, such as Dye 3 and Dome C, the site elevation during the Late Wisconsin is not expected to differ much from the present elevation. Indeed, the air volume data at Dye 3 indicate no substantial elevation change (S. Herron, personal communication, 1982). Therefore Cl^- concentration variations at those sites should mostly reflect changes in source activity, providing the scale height is largely unaffected.

At sites such as Camp Century and, particularly Byrd Station, Antarctica, however, it is likely that deep ice fell as snow at a site elevation considerably different from the present site elevation. If only two time periods are considered, Holocene Cl^- concentrations would be described by

$$C_H = C_0 \exp\left(-\frac{Z_H}{1.52 \text{ km}}\right), \quad (3)$$

analogous to equation (1) with Z_H representing the current site elevation. For Late Wisconsin concentrations,

$$C_W = C_1 \exp\left(-\frac{Z_W}{1.52 \text{ km}}\right), \quad (4)$$

where C_1 reflects the increased sea-salt aerosol production during the Late Wisconsin and Z_W is the elevation of origin of the Late Wisconsin ice. Combining equations (3) and (4) yields the elevation difference, ΔZ:

$$\Delta Z = Z_W - Z_H = 1.52 \text{ km} \cdot \ln\left(\frac{C_1}{C_0} \cdot \frac{C_H}{C_W}\right) \quad (5)$$

The Dye 3 and Dome C results suggest that C_1/C_0 was about 4–5. The limited Camp Century Cl^- data yield a C_H of 35 ng/g and C_W of 120 ng/g. From these results, ΔZ is calculated to be 400 m, in accord with the air volume data of S. Herron [1982] and with the conclusions of Paterson [1977] who indicate an elevation difference of 400–500 m. This agreement should be regarded as fortuitous until more abundant chemical data are available.

Chloride data are not available for the Byrd Station core, but limited Na data are available [Cragin et al., 1977] and Na is a suitable index element for sea salts in Antarctica. Average Holocene concentrations at Byrd are 28 ng/g while during the Late Wisconsin the average was 48 ng/g. Using the Dome C values for the C_1/C_0 ratio of 3.8–4.8, these data suggest that Late Wisconsin ice was formed 1.2–1.6 km higher than the present Byrd Station elevation of 1520 m. Such high ice elevations are not found in West Antarctica today, but the estimated elevation difference is consistent with the reconstructed West Antarctic ice sheet dimensions for the Late Wisconsin [Stuiver et al., 1981].

These examples outline the potential use of equation (5) for providing important paleoenvironmental information from the Cl^- concentration profiles of deep ice cores. This technique and the underlying assumptions are not established, but may provide a direction for future glaciochemical research.

Sulfate and Accumulation Rates

Sulfate in pre-industrial polar snow is derived from three sources: (1) biogenic sources that produce a stable background SO_4^{2-} concentration, (2) volcanic eruptions that produce episodic SO_4^{2-} peaks followed by SO_4^{2-} concentration which rapidly decline to the background level, and (3) sea salts, which are generally an insignificant source at non-coastal sites, but are important during the Late Wisconsin. The sea-salt SO_4^{2-} component is assumed to be equal to $0.14 \times Cl^-$, where 0.14 is the SO_4^{2-}/Cl^- ratio in bulk seawater. Distinguishing the background SO_4^{2-} concentration from possible volcanogenic contributions requires the detailed sampling and analysis that has been done on the Dye 3 core.

The background SO_4^{2-} concentration is important to establish since this component shows a strong dependence on the snow accumulation rate, A. In Holocene samples from a number of Greenland and Antarctic sites, the background SO_4^{2-} concentration, S_B, may be expressed as

$$S_B = S_0 A^{-1} \quad (6)$$

where S_0 is the background sulfate flux [M. Herron, 1982]. S_0 is a constant with Greenland values about twice those measured in Antarctica. The widespread applicability of equation (6) over large regions of Greenland and Antarctica suggests that it may be possible to derive past snow-accumulation rates from SO_4^{2-} concentration profiles. Utilization of equation (6) for that purpose requires that the background SO_4^{2-} flux be constant or that past variations in flux be known.

Antarctic SO_4^{2-} profiles indicate that the flux of background SO_4^{2-} during the Wisconsin may have been quite similar to the flux today. The Dome C SO_4^{2-} concentration profile has not yet been published but is described qualitatively as showing no significant changes during the Wisconsin [Petit et al., 1981]. The accumulation rate record determined by microparticle stratigraphy also shows no significant change [Thompson et al., 1981]. The Dome C chronology of Lorius et al. [1979] indicates that snow-accumulation rates were not less than 50% of present values. Within these qualitative restrictions the Dome C SO_4^{2-} flux appears at least not greatly changed from Holocene levels.

Sulfate concentrations have been published for a few depth intervals of the Byrd Station deep ice core [Cragin et al., 1977]. Excluding a single sporadic data point that may include a volcanogenic contribution, the average SO_4^{2-} concentration in Late Wisconsin ice is 74 ng/g compared to the Holocene average of 61 ng/g. Corrected for the sea-salt contribution, the Late Wisconsin/Holocene concentration ratio is 62/54 = 1.15. These results suggest that the snow accumulation rate was 87% of the Holocene rate if the background flux was constant. In view of the Dome C results, the Late Wisconsin background SO_4^{2-} flux was most likely not very different from the present flux.

In contrast to the situation in Antarctica, Late Wisconsin background SO_4^{2-} concentrations in Greenland are markedly higher than Holocene concentrations. If equation (6) was also valid during the ice age, and if, as the Antarctic results suggest, the background SO_4^{2-} flux was unchanged, the SO_4^{2-} concentration profiles in Greenland can be translated into snow accumulation rates.

Table 1 gives background SO_4^{2-} concentrations and derived paleoaccumulation rates for several intervals of the Dye 3 core. The background SO_4^{2-} concentrations have had the sea-salt component subtracted using the Cl^- concentrations measured on the same samples. With the abundance of Dye 3 data, the background level may be estimated fairly easily, even during the Late Wisconsin when there was apparently substantial volcanic activity. This is not the case for the Camp Century core, though the SO_4^{2-} levels appear to generally match the features of the Dye 3 core. The derived accumulation rates in Table 1 indicate that snow accumulation rates during the Late Wisconsin were as little as 22% of the present rate of 490 kg $m^{-2}a^{-1}$e. Further, the derived accumulation rates never exceed the Holocene rate for the entire period encompassed by the deep core.

The implications of the results in Table 1 are far-reaching. They suggest that age models based on a constant accumulation rate will significantly underestimate the age and overestimate the annual layer thickness of pre-Holocene ice. It may be noted in this regard that annual layer thicknesses measured on Late Wisconsin ice from Camp Century [Hammer et al. 1978] are generally much lower than estimated from the Hammer et al. [1978] model. Annual layer thicknesses will, however, be difficult to estimate from the data in Table 1 since the rate of thinning is partially dependent on the driving force provided by new snowfall. Therefore the rate of thinning of deep ice was probably lower in the past than at present. These results suggest that accurate modeling of the depth-age relationship in pre-Holocene ice from Greenland will require non-steady state assumptions.

Sulfate and Volcanism

One of the aims of SO_4^{2-} analyses on polar ice cores is to obtain the history of explosive volcanism. Several examples of the impact of historic and prehistoric eruptions on the composition of Greenland precipitation were given by M. Herron [1982]. The multi-year impact of numerous Holocene- and Wisconsin-age eruptions is demonstrated in the Dye 3 SO_4^{2-} peaks in Figures 2, 4, and 6. The peaks just below the 1786 m transition could conceivably correspond to the eruption of Glacier Peak, Washington, (11.2 kBP) or the Eiffel eruptions of 10.0–12.0 kBP which were the last European eruptions north of the Alps [Lamb, 1971]. However, in view of the extremely incomplete chronology of explosive volcanism and the uncertainty of the ice core depth-age relationship, such correlations are very tentative.

The Late Wisconsin SO_4^{2-} peaks are extremely large in comparison with Holocene events. For example, the 1766 AD eruption of Mayon, Luzon, is estimated to be the third largest in terms of atmospheric dust within the past 300 years [Lamb, 1971]. The Dye 3 SO_4^{2-} concentrations for the three years following the eruption averaged 57 ng/g, about 34 ng/g above the sea-salt and biogenic baseline. For the three years following the 934 AD eruption of Eldgja, Iceland, the SO_4^{2-} concentrations in nearby Dye 3 averaged 65 ng/g above background. In comparison to these Holocene events, the SO_4^{2-} peak at 1789.3 m in the Dye 3 core is at least 80 ng/g above the Late Wisconsin baseline. This sample is estimated to represent about three years according to the preliminary Dye 3 age model. If, as suggested above, annual-layer thicknesses are considerably smaller, the number of years represented would be greater and the volcanic impact even more severe.

The peak at 1789.3 m is only one of numerous large SO_4^{2-} peaks in the Dye 3 core that suggest that volcanic activity was significantly higher during the Late Wisconsin than it is today. In comparing the Wisconsin and Holocene SO_4^{2-} peaks in Figure 2, it must be emphasized that most of the Holocene volcanic intervals were not chosen at random but were selected because of elevated solid conductivity values [Hammer et al., 1984]. In contrast, Wisconsin ice samples were selected more or less at random because the solid conductivity technique does not work in Wisconsin-age ice from Greenland. Thus, the relative level of Holocene volcanism versus Wisconsin volcanism is somewhat overestimated in the SO_4^{2-} profile in Figures 2 and 6.

According to the preliminary time scale, the large SO_4^{2-} peaks represent the periods in thousands of years BP: 10.4–15, 16.6, 17.8, 19.0, 22.5, 27.5, 28.5, 29.6, 30.0, 30.2, and 31.0. In addition, small SO_4^{2-} peaks at 54.2 and 57.0 kBP may include a significant volcanic input that is greatly reduced due to the large number of years represented by each sample.

Petit et al. [1981] suggest that there is no evidence of Late Wisconsin volcanism from the microparticle analysis of the Dome C core. The Dye 3 results suggest that evidence of Late Wisconsin volcanism will be discernible from the SO_4^{2-} concentration profile. If the above interpretations are correct, then it is predicted that in the Dome C core the Late Wisconsin background SO_4^{2-} concentrations will be up to 50% greater than the Holocene background, and that numerous SO_4^{2-} peaks, reflecting large scale volcanic activity in the latter part of the ice age, will be superimposed on this background.

Nitrate

Interpretation of the long-term changes in NO_3^- concentrations in the Dye 3 and Camp Century ice cores is hampered by a lack of knowledge concerning the sources of No_3^- in polar precipitation and the parameters that control NO_3^- concentration variations. Inasmuch as the source(s) of NO_3^- are still unknown, estimates of the effects of ice age conditions on source activity are questionable.

For most, but not all polar sites, NO_3^- concentrations show a dependence on snow-accumulation rate:

$$N = N_0 A^{-1/2} \quad (7)$$

where N is the NO_3^- concentration in pre-industrial precipitation, A is the snow-accumulation rate, and N_0 is a constant with values in Greenland about twice those in Antarctica [M. Herron, 1982]. Following equation (7), Dye 3 NO_3^- concentration variations over the 1786 m transition indicate accumulation rate changes equal to those derived from the background SO_4^{2-} results. The average NO_3^- concentration at Dye 3 over the 1786.1–1788 m depth interval is 74 ng/g. Above 1786.1 m, the average concentration is 51 ng/g, indicating a change in accumulation rate of $(74/51)^2$ or 2.1. This is equivalent to the SO_4^{2-} derived accumulation rate change over this interval (Table 1).

Below the transition, Dye 3 NO_3^- concentrations decrease with increasing depth, reaching minimum concentrations around 1835 m of 35–40 ng/g, lower than average concentrations in the Holocene. Late Wisconsin nitrate concentrations are also below Holocene average in the Camp Century core, and NO_3^- is the only chemical constituent yet measured to show this relationship.

One tentative interpretation of the Greenland NO_3^- profiles is that while lower snow-accumulation rates and perhaps increased meridional circulation during the Late Wisconsin would be expected to produce elevated concentrations, the observed low concentration levels reflect a substantial decline in NO_3^- source activity. A preliminary comparison with the Dye 3 $\delta(^{18}O)$ curve of Dansgaard et al. [1982] indicates that below 1835 m, the NO_3 and δ curves correlate fairly well, with low NO_3^- concentrations accompanying more negative (colder) δ-values. Above 1835 m the correlation is lost and, at the 1786 m transition, the correlation is reversed with higher NO_3^- concentrations matching colder δ-values. These results suggest the possibility that more than one source is responsible for NO_3^- in Greenland snow. One source, dominant in ice deeper than 1835 m, is apparently more active during warmer conditions. The second source, operative in ice above 1835 m and dominant near the transition, apparently follows the accumulation rate dependence described by equation (7) which holds for most polar sites today.

Summary and Conclusions

The anion concentration profiles of the Dye 3 and Camp Century ice cores have shed new light on the climatic and environmental changes of the past. Chloride concentrations at Dye 3 and other stable sites indicate that high wind speeds during the Wisconsin glacial stage increased sea-salt aerosol concentrations over the ocean and resulted in Cl^- concentrations as much as 4–5 times the levels in recent times. The increased ice age wind speeds are also thought to be a major factor in producing high concentrations of eolian dust during the Late Wisconsin. The magnitude of the sea-salt concentration increase is less at rheologically unstable sites such as Camp Century and, in particular, Byrd Station in West Antarctica. Chloride concentration profiles have been interpreted in light of their present dependence on site elevation to indicate that Wisconsin-age ice at Camp Century originated approximately 400 m higher than the present elevation. Sea-salt concentrations in the Byrd core indicate that Late Wisconsin ice could have formed at an elevation of 2700–3100 m. This is higher than any elevation in the existing ice sheet, but is consistent with an expanded West Antarctic ice sheet envisioned in Late Wisconsin reconstructions.

Variations in background SO_4^{2-} concentrations at Dye 3 and Camp Century indicate that the snow-accumulation rate during the Wisconsin may have been substantially lower than the present rate. During the driest period, in the Late Wisconsin, the accumulation rates were apparently as low as 22% of the modern rate. These results suggest that there may be substantial errors in pre-Holocene ages determined from models based on steady-state assumptions. The low values might also indicate a starvation of the Late Wisconsin ice sheets which may have been a contributing factor in their demise. In contrast to the situation in Greenland, the Byrd Station SO_4^{2-} profile indicates that Antarctic snow-accumulation rates were only slightly lower during the Wisconsin than the present.

In addition to being characterized by high-wind speeds and reduced-precipitation rates, the Late Wisconsin was also a time of extremely high-volcanic activity. This interpretation is based on the large SO_4^{2-} concentration peaks in Late Wisconsin ice from Dye 3. Several of the concentration perturbations are larger than any yet measured in Dye 3 Holocene ice deposited after historic or prehistoric eruptions. Given the pronounced climatic impact of such eruptions as Krakatoa, the Dye 3 results suggest that the Late Wisconsin cold period may have been intensified or extended as a result of the high level of explosive volcanism.

Nitrate concentration variations in the Dye 3 and Camp Century ice cores will remain an enigma until more information is available regarding sources and parameters that govern observed concentration. Nitrate is the only measured chemical constituent with concentrations in Late Wisconsin lower than in Holocene ice. Below 1835 m in the Dye 3 core, NO_3^- variations largely parallel the $\delta(^{18}O)$ changes indicating a possible temperature dependence on source activity. However, for samples at and above the 1786-m transition, NO_3^- concentrations appear to follow variations in snow-accumulation rate, as found in present-day snow from various locations. These results may indicate that there is more than one important source of NO_3^- in Greenland precipitation.

Finally, the incredible rapidity of the chemical transition at 1786 m should be emphasized. Within tens of years, the chemical signal of Dye 3 ice changed from an ice age character to a Holocene character. The implications of the Cl^-, SO_4^{2-}, and NO_3^- curves are that within perhaps as little as a decade the high winds over the oceans dropped to Holocene levels and the snow accumulation rate increased by more than a factor of two. The suddenness of this transition, if confirmed by detailed chemical analysis of other ice cores, indicates that the time scale for major climatic change may be of the order of decades.

Acknowledgments. We thank S. Ribes, R. Ruszczyk, and S. Rumer for field and laboratory participation. This work was sponsored by the National Science Foundation, Division of Polar Programs.

References

Cragin, J. H., M. M. Herron, C. C. Langway, Jr., and G. A. Klouda, Interhemispheric comparison of changes in the composition of atmospheric precipitation during the Late Cenozoic Era, in *Polar Oceans*, edited by M. J. Dunbar, pp. 617–631, Arctic Institute of North America, Calgary, Alberta, 1977.

Dansgaard, W., and N. Reeh, Climatic interpretation of GISP ice core data (abstract), *EOS Transactions, American Geophysical Union, 63*, 298, 1982.

Dansgaard, W., H. B. Clausen, N. Gundestrup, C. U. Hammer, S. F. Johnsen, P. M. Kristinsdottir, and N. Reeh, A new Greenland deep ice core, *Science, 218*, 1273–1277, 1982.

Hammer, C. U., H. B. Clausen, W. Dansgaard, N. Gundestrup, S. J. Johnsen, and N. Reeh, Dating of Greenland ice cores by flow models, isotopes, volcanic debris, and continental dust, *Journal of Glaciology, 20*, 3–26, 1978.

Herron, M. M., The impact of volcanism on the chemical composition of Greenland Ice Sheet precipitation, Ph.D. thesis, 158 pp., State University of New York at Buffalo, 1980.

Herron, M. M., Impurity sources of F^-, Cl^-, NO_3^-, and SO_4^{2-} in Greenland and Antarctic precipitation, *Journal of Geophysical Research, 87*, 3052–3060, 1982.

Herron, S. L., Physical properties of the deep ice core from Camp Century, Greenland, Ph.D. thesis, 153 pp., State University of New York at Buffalo, 1982.

Lamb, H. H., Volcanic activity and climate, *Paleogeography, Paleoclimatology, Paleoecology, 10*, 203–230, 1971.

Langway, C. C., Jr., Stratigraphic analysis of a deep ice core from Greenland, *Geological Society of America Special Paper, 125*, 186 pp., 1970.

Lorius, C., L. Merlivat, J. Jouzel and M. Pourchet, A 30,000 year isotope climatic record from Antarctic ice, *Nature, 280*, 644–648, 1979.

Murozumi, M., T. J. Chow and C. Patterson, Chemical concentrations of pollutant lead aerosols, terrestrial dusts and sea salts in Greenland and Antarctic snow strata, *Geochimica et Cosmochimica Acta, 33*, 1247–1294, 1969.

Paterson, W. S. B., Extent of the late-Wisconsin in glaciation in northwest Greenland and northern Ellesmere Island: A review of glaciological and geological evidence, *Quaternary Research, 8*, 180–190, 1977.

Petit, J. R., M. Briat and A. Royer, Ice age aerosol content from East Antarctic ice core samples and past wind strength, *Nature, 293*, 391–394, 1981.

Ruddiman, W. F. and A. McIntyre, Late Quaternary surface ocean kinematics and climatic change in the high-latitude North Atlantic, *Journal of Geophysical Research, 82*, 3877–3887, 1977.

Stuiver, M., G. H. Denton, T. J. Hughes and J. L. Fastook, History of the marine ice sheet in West Antarctica during the last glaciation: A working hypothesis, in *The Last Great Ice Sheets*, edited by G. H. Denton and T. J. Hughes, pp. 319–436, John Wiley and Sons, New York, 1981.

Thompson, L. G., E. Mosley-Thompson and J. R. Petit, Glaciological interpretation of microparticle concentrations from the French 905-m Dome C, Antarctica core, in *Sea Level, Ice, and Climatic Change*, edited by I. Allison, pp. 227–234, *IAHS-AISH Publication No. 131*, 1981.

CO_2 CONCENTRATION IN AIR EXTRACTED FROM GREENLAND ICE SAMPLES

B. Stauffer, A. Neftel, H. Oeschger, and J. Schwander[1]

Physics Institute, University of Bern, Switzerland

Abstract. The principal aim of the analyses of the CO_2 concentration in air extracted from ice samples is to reconstruct the CO_2 concentration of the atmosphere during the last millenia. For this purpose ice from very cold regions is best suited. Ice samples from Dye 3, where the mean annual air temperature is −20°C and summer melting is frequent, are not very well suited from this point of view. The results of CO_2 analyses give however very valuable information on a possible temperature effect on the CO_2 concentration of air in the bubbles. The CO_2 content shows seasonal variations with an annual maximum value in the summer melt layer. The annual minimum values correspond approximately to the estimated atmospheric CO_2 concentrations. Based on this experience, in spite of the complications due to the melt features, we try to reconstruct the history of the CO_2 concentrations of the atmosphere. Of special interest are the fast climatic transitions in the course and especially at the end of the last glaciation which are represented in the ice core by changes of parameters like acidity, dust, and istopic ratios in short depth intervals.

In this respect we discuss the time lag between the climatic warming at the end of the last glaciation and the increase of the atmospheric CO_2 concentration. Finally results of gas content and gas composition of two ice samples from the lowest, silty part of the ice core are discussed.

Introduction

The principal aim of the analyses of the CO_2 concentration in air extracted from ice samples is to reconstruct the CO_2 concentration of the atmosphere during the last millenia. For this purpose ice from very cold regions is best suited, since any interaction of air with water, which leads to a change in the gas composition, can be excluded. Ice samples from Dye 3, where the mean annual air temperature is −20°C and summer melting is frequent, are, therefore, not ideal for direct information on the history of the atmospheric CO_2 concentration, but offer unique possibility to study the influence of ice temperature and melt layers on the CO_2 concentration in air bubbles. An additional problem regarding reconstruction of ancient atmospheric CO_2 concentrations is a possible interaction of air with the ice structure.

At Dye 3, below 1300 m depth, air diffuses from bubbles into the ice structure [Shoji and Langway, 1982]. To study whether this diffusion influences the amount and composition of the gases extracted by a dry or melt extraction method, we analyzed ice samples a few days, a few weeks, and a few months after recovery. A change in the gas composition can hardly be detected and the results, which are discussed below, are not affected by this diffusion phenomenon. Details of these studies are discussed in another paper [Neftel et al., 1983].

From analyses of ice samples from Camp Century and Byrd Station, there is strong evidence that the CO_2 concentration of the atmosphere was considerably lower during the last part of the Last Glaciation than in the Holocene [Neftel et al., 1982]. In spite of the difficulties due to the melt layers, the analyses of the samples from Dye 3 provide important new information concerning such a transition from lower to higher atmospheric CO_2 concentrations at the Glacial-Postglacial transition.

Origin of Ice Samples and Experimental Procedure

The ice samples, the results of their analysis we will discuss, were recovered during the GISP deep drilling operation at Dye 3 in 1981 [Gundestrup and Johnsen, 1985]. A few samples were transported within two days after recovery to the laboratory in Bern, enabling us to analyse the gas in freshly recovered ice cores. The firn samples were obtained by core drilling with the Swiss electromechanical shallow drill in 1980.

Gas analysis is performed on samples of 300 g ice using a melt extraction technique and subsequent gas composition measurement with a gas chromatograph, and on samples of 1 g ice using a dry extraction method and determination of the CO_2 concentration with a laser absorption spectrometer. The samples are cut to the required size from the inner part of the core with a band saw. The 300 g samples, in addition, are cleaned with a microtome knife on all surfaces. They are afterwards melted in an evacuated glass container. The escaping gases are extracted continuously, and the gas samples are measured in different fractions in a gas chromatograph Hewlett Packard 5880 A. N_2, O_2, and Ar are measured with a thermal conductivity detector. CO_2 is catalytically transformed into methane and then measured with a flame ionization detector [Oeschger et al., 1982]. The accuracy for an analysis of N_2, O_2, and Ar is about 0.5% rel. The overall accuracy of the CO_2 concentration measurement of the extracted air is about 5% rel. (15 ppm abs.). The air extracted by melting also contains CO_2 originating from carbonates in the ice [Stauffer et al., 1982].

Ice cubes of 1 g are crushed at −20°C in a specially designed crusher. The escaping air from the opened bubbles expands into an evacuated absorption cell of the laser spectrometer. The CO_2 concentration is determined by measuring the absorption of a beam from a tunable diode laser at 4.3 μm wave length. The overall accuracy using 1 g samples is about 2% rel. (6 ppm abs.). More details concerning this method are given by Zumbrunn et al. [1982].

[1] Now at: State University of New York at Buffalo, Department of Geological Sciences, Amherst, New York.

TABLE 1. Solubilities and Diffusion Constants of Air Components in Water.

	air	N_2	O_2	Ar	CO_2
Composition of atmospheric air	100%	78.08%	20.95%	0.93%	300 ppm
Air dissolved in water at 0°C and atmospheric pressure	100% (29.5 ml/kg)	61.6%	34.7%	1.77%	19 300 ppm
Diffusion constant (m^2s^{-1})	—	$1.61 \cdot 10^{-9}$	$208 \cdot 10^{-9}$	$1.26 \cdot 10^{-9}$	$1.18 \cdot 10^{-9}$

Seasonal Variations of the Gas Content and the Gas Composition of Ice Samples from Dye 3

At Dye 3, almost every summer during short periods, the surface air temperature rises above 0°C and melt layers are formed. The contribution of melt layers to the total ice at Dye 3 under today's climatic conditions is about 7% by volume [Herron et al., 1982]. The gas content and composition in melt layers is completely different from those of ice formed by sintering of dry firn.

If melt water saturated with air penetrates into the cold it refreezes as a thin layer of ice covering cold snow grains, as ice grains caught between cold grains or, only in case of very heavy melt-water production, as a compact ice layer above an impermeable firn layer. During refreezing, part of the dissolved gases escape by diffusion. Since for the different air components the diffusion constants are different, the gas content and the gas composition in frozen melt water are not the same as in liquid water saturated with air. Solubilities and diffusion constants for the main components of air are given in Table 1. O_2 has a high diffusion constant and is therefore expected to be depleted in a frozen melt layer relative to the other gases when compared to gas concentrations in air-saturated water.

The amount of the composition of the gases enclosed during refreezing depends mainly on the freezing speed. Melt layers are only seldomly compact layers of ice, but more frequently ice grains and ice lenses that take part in the sintering process of the firn. At the transition from firn to ice, additional air with atmospheric composition is also enclosed in melt-layer ice. The enclosed amount varies, and is generally smaller than the amount which is enclosed in ice sintered from dry firn. Based on measurements by Herron et al. [1982], we estimate an average of 60 ml air with atmospheric composition per kg melt-layer ice, compared to 100 ml per kg ice if sintered from dry firn. We measured the gas content and composition of different melt layers from the firn at Dye 3 originating from depths between 1 m and 35 m below the surface. Since at this shallow depth no pores are sealed off yet due to sintering, these measurements reveal only the amount and composition of air enclosed during the refreezing process. The results are given in Table 2. For all samples very high CO_2 concentrations are observed. The first two samples were collected during the excavation of a large trench at Dye 3. They are the largest and most compact ice pieces and, therefore, these measurements represent probably extreme values. The other five samples were taken from a shallow firn core and are considered to be more representative of typical melt-layer ice.

The mean CO_2 concentration in air bubbles thus depends strongly on the presence of melt layers. But, since the CO_2 surplus in melt layers shows large variations, it is not possible to calculate quantitatively the melt layer contribution from CO_2 concentration measurements. A very rough estimate can be made, however, using the average values. Melt-layer ice contains about 67 ml of air per kg ice, with a CO_2 concentration of about 1500 ppm. This value is obtained by assuming that it contains 60 ml of air with atmospheric composition, and about 7 ml of air with a CO_2 concentration of about 12,000 ppm which is the average of our measurements as shown in Table 2. It is further assumed that part of the ice that is sintered from dry firn contains 100 ml of air with atmospheric composition. The roughly estimated melt-layer percentage as a function of the CO_2 concentration is shown in Table 3. In Figure 1, CO_2 concentration measurements on ice samples from 124 m depth below surface are shown. The two samples that contain melt-layer ice show an increased CO_2 concentration as expected. The mean CO_2 concentration for the eleven samples is 660 ppm. According to Table 3, the estimated melt-layer contribution would be 40% which is obviously an overestimation by about a factor of 4. This comparison shows clearly the limits of our estimates for single-melt layers. In samples without melt layers, the CO_2 concentration is between 280 ppm and 330 ppm. The dispersion of the results is greater than in samples from colder locations and the mean concentration (315 ppm) is higher than that expected for the preindustrial atmospheric CO_2 concentration. Measurements of ice samples from colder locations indicate this concentration to be about 270 ppm [Oeschger et al., 1982]. Based on the Dye 3 data, in ice formed by sintering of dry but relatively warm firn (above −30°C), CO_2 seems to be slightly enriched.

In Figure 2, CO_2 concentrations from ice samples from 1616 m are shown. High CO_2 concentrations are no longer confined to narrow

TABLE 2. Content and Composition of Air Extracted From Melt Layers.

Depth below surface (%)	Total gas content (ml kg^{-1})	N_2 (%)	O_2 (%)	Ar (%)	CO_2 (ppm)
	14.5	63.0	32.5	1.65	28 000
	13.4	68.4	28.5	1.48	16 100
0.7	4.2	73.64	23.9	1.14	10 700
6.9	3.35	75.62	22.74	1.05	4 070
11.3	7.47	70.86	26.73	1.28	10 600
28.5	2.48	75.01	22.73	1.12	9 400
34.5	3.06	75.36	22.65	1.09	7 400
average:	6.92	71.70	25.68	1.26	12.324

TABLE 3. Estimated Air Content and Composition as a Function of Melt-Layer Contribution.

Melt layer contribution (%)	Total gas content (ml kg^{-1})	N_2 (%)	O_2 (%)	Ar (%)	CO_2 (ppm)
0	100	78.08	20.95	0.93	300
5	98.35	78.05	20.97	0.93	343
10	96.69	78.03	20.98	0.93	386
15	95.04	78.01	21.00	0.93	432
20	93.38	77.99	21.02	0.93	478
30	90.08	77.93	21.06	0.94	578
40	86.77	77.88	21.10	0.94	684

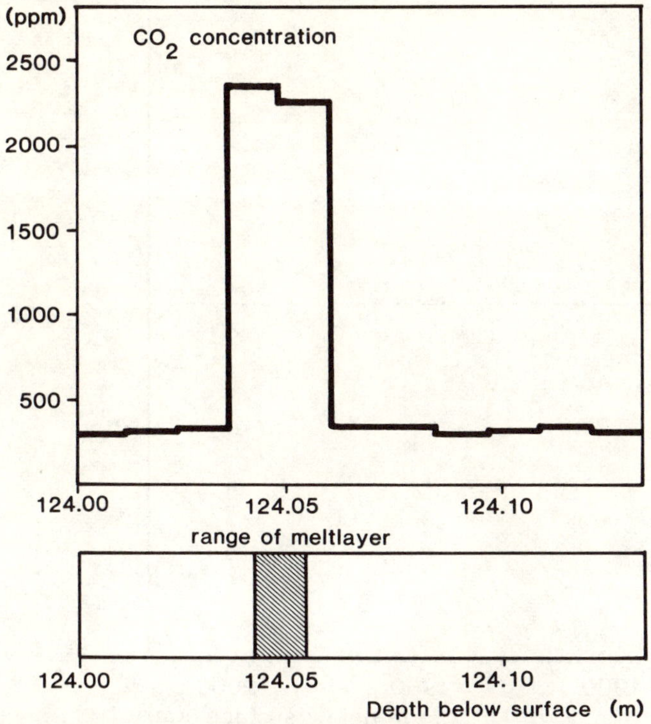

Fig. 1. CO_2 concentration of air extracted from ice samples from a 0.12 m long core from Dye 3 including one melt layer.

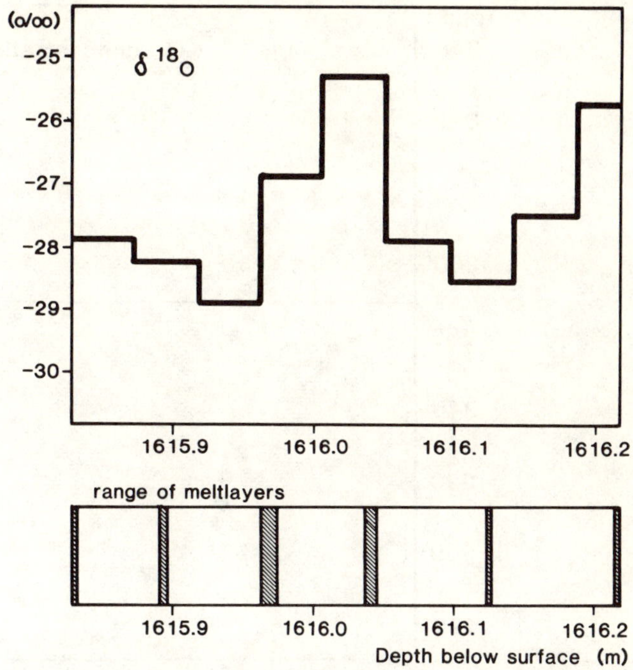

Fig. 3. Estimated initial distribution of the melt layers based on CO_2 concentration measurements.

melt layers as in the samples from 124 m depth. Obviously, during the time the ice needs to reach that depth, CO_2 diffused from the melt layers into the ice formed by sintering of dry firn. Diffusion calculation based on these data show that the diffusion constant for CO_2 in polycristalline ice is in the order of the constant for self

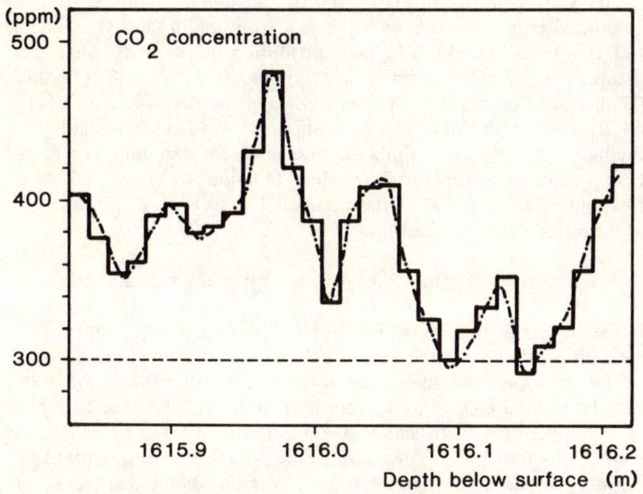

Fig. 2. CO_2 concentration of air extracted from ice samples from a 0.39 m long ice core. The dashed line was calculated by assuming that the melt layers had a distribution as shown in Figure 3, and had at shallow depth all average CO_2 concentration.

diffusion ($6 \cdot 10^{-16}$ m^2 at $-20°C$) [Neftel et al., 1983]. The mean CO_2 concentration at this depth is 375 ppm. Based on Table 3, it is estimated that the melt layer contribution in this ice is about 9%. Assuming that each maximum of the CO_2 concentration corresponds to a summer melt layer and therefore separates annual layers, the CO_2 concentration for each annual layer can be estimated. The result of such an estimate is shown in Figure 3. It has to be kept in mind that the estimates of single melt layers can be off by a factor of 4 as shown above. Also in Figure 3, $\delta^{18}O$ results measured by Dansgaard et al. (personal communication) are shown. The resolution of the measurements is not high enough to allow seasonal variations to be seen. However, it is of interest to observe that the highest $\delta^{18}O$ values are found for the periods with the thickest ice layers, as one would expect if higher mean $\delta^{18}O$ values also indicated elevated summer temperatures. Based on the information we have at present, we conclude that it is reasonable to assume that high CO_2 concentrations reflect ice with a relatively high melt-layer contribution. A quantitative estimate, however, seems difficult but it should be possible to detect general trends; e.g. warmer summers during the climatic optimum. CO_2 concentration measurements would allow also to count annual layers at least down to a depth of 1600 m, corresponding to about 7000 yrs B.P.. For this purpose, however, simpler methods as, e.g., the measurements of the electric conductivity are already available. A very important result regarding the interpretation of CO_2 measurements on ice cores is the small diffusion rate of CO_2 in polycristalline ice.

Long-Term Variations of the CO_2 Content

Long-term variations of the CO_2 concentration measured in ice from Dye 3 may either be due to a change in the average annual melt-layer thickness, or due to a variation of the atmospheric CO_2 concentration. Since melt layers can only lead to CO_2 concentra-

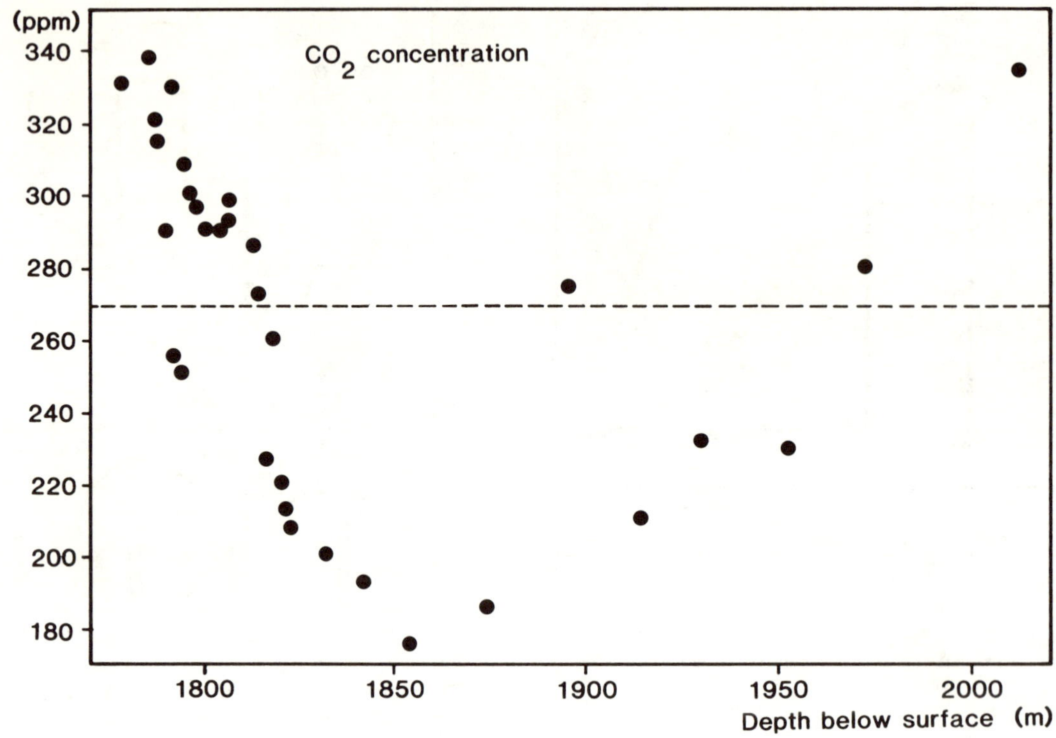

Fig. 4. CO_2 concentration of air extracted from 1 g ice samples from the Dye 3 deep ice core.

tions that are higher than the atmospheric one, the measurements give at least an upper limit of the atmospheric CO_2 concentration.

In Figure 4 results are shown from measurements of the CO_2 concentration of air extracted from ice samples between 1775 m and 2020 m depth. Between 1820 and 1950 m, with one exception, values that are significantly below the estimated preindustrial atmospheric CO_2 concentration of about 270 ppm are observed. Low values are found already at 1792 m depth, but the pronounced shift to low values occurs at 1815 m below surface. According to a tentative time scale, this depth corresponds to an age of 11,300 yrs B.P. According to another estimate [Oeschger, 1985], it may correspond to 13,000 yrs B.P. (^{14}C-time scale). This change from high to low CO_2 values occurs 30 m deeper than the pronounced change of several physical and chemical parameters observed at 1786 m below surface, indicating the end of Last Glaciation (10,000 yrs B.P.). There are two possible explanations for this phase difference. Either the atmospheric CO_2 concentration increased before the final termination of the glaciation in Greenland, or there were warm summers toward the end of the glacial epoch with surface melting leading to CO_2 enrichments.

The $\delta^{18}O$ values in the depth interval from 1775 m to 2020 m are all below −30.5‰ [Dansgaard et al., 1982]. Under today's climatic conditions this corresponds to a mean annual air temperature of below −24.5°C, and at this temperature summer melting should be negligible. It is, however, theoretically not allowable to derive in such a simple way from $\delta^{18}O$ values the mean annual air temperatures in another climatic epoch. And even in the case where the mean annual air temperature had been low, it still could have been possible that there was summer melting, since according to Milankovitch the summer insolation in the northern hemisphere was at a maximum during that time [Broecker and van Donk, 1970]. From the results on the Dye 3 ice samples alone it is difficult to decide whether the CO_2 shift at 1815 m depth reflects an atmospheric CO_2 concentration change, or a melt-layer artifact. The depth of 1815 m probably corresponds to the period of the first warming (Oldest Dryas-Bølling transition) as observed, e.g., in European peat bog pollen profiles and lake sediment $\delta^{18}O$ profiles. If we exclude a dominating melt layer effect, this warming was accompanied by a significant atmospheric CO_2 increase influencing the terrestrial radiation balance. Another explanation would be a slow, steady atmospheric CO_2 increase as shown in Figure 5 with superimposed higher CO_2 concentrations of the air occluded in the ice due to summer melting, starting with the first warming in the Late Glacial. For this model the melt layer contribution can be estimated. The estimate is given in Figure 5 and compared with the $\delta^{18}O$ values [Dansgaard et al., 1982]. A comparison of the necessary melt-layer contribution with the $\delta^{18}O$ values shows that it is very unlikely that the fast CO_2 increase is a melt layer artifact since no melt layers can be expected in Greenland ice with $\delta^{18}O$ values as low as −36‰. A fast and early increase of the atmospheric CO_2 concentration is a more probable interpretation.

Gas Content and Composition in the Silty Ice Layer

The lowest 22 m of ice at Dye 3 are silty. In Table 4, results of the measurements of two ice samples from this silty layer are shown.

The total gas content is only about two thirds of that above the silty layer, and CO_2 is extremely enriched, probably due to a high carbonate content. The gas was extracted with the melt extraction method so that carbonates contribute to the CO_2 content. Ar is slightly enriched, but O_2 is depleted. The same characteristics of gas content and gas composition were found in the silty layer at the bottom of the ice sheet at Camp Century [Herron and Langway, 1979]. The reason for the lower gas content, the enrichment of Ar and the depletion of O_2 compared to N_2 is thought to be a diffusion process. Air may have diffused from ice with ordinary gas content

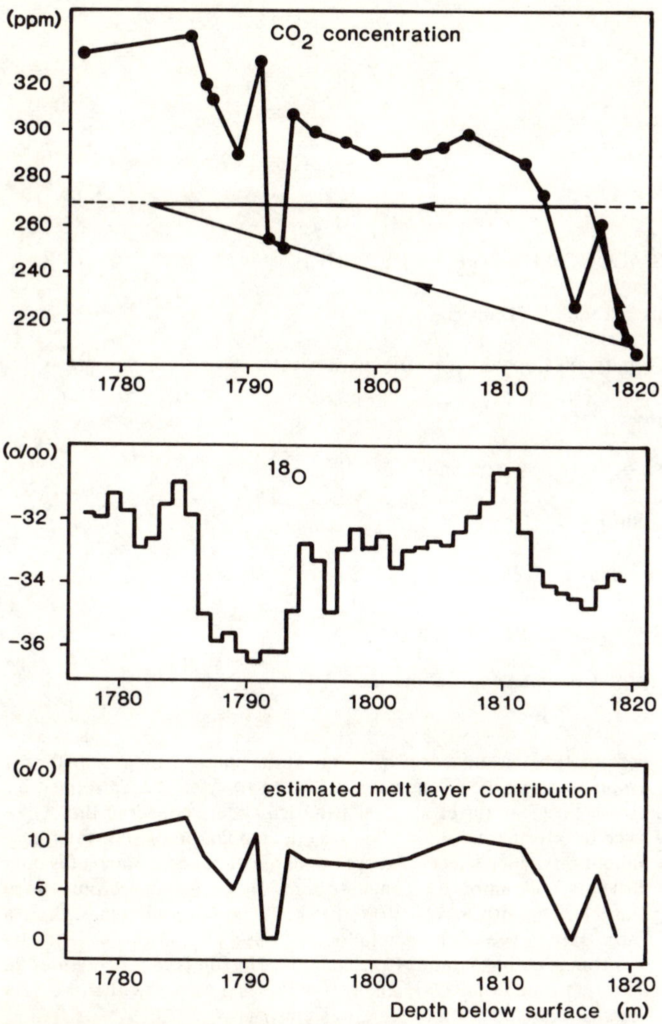

Fig. 5. Estimated melt-layer contribution assuming that atmospheric CO_2 concentration increased according to line b. A comparison with $\delta^{18}O$ values given by Dansgaard et al. [1982], shows that the formation of melt layers is unlikely and that the atmospheric CO_2 concentration slightly increased according to line a.

and composition into a layer of refrozen melt water with very low gas content at the bottom of the silty ice layer, at a time when the silty ice had been at the pressure melting point. The temperature at the bottom of the ice sheet at Dye 3 have not yet been measured with the desired accuracy, but from thermometers in the drill we know that it is of the same order as at Camp Century (−13°C) [Weertman, 1968]. Therefore the results of the gas measurements may be interpreted similarly. Dansgaard et al. [1982] have measured in the silty ice at Camp Century low and varying $\delta^{18}O$ values but in Dye 3

TABLE 4. Air Content and Composition of Samples of Silty Ice.

Depth below surface (m)	total gas (ml/kg)	N_2 (%)	O_2 (%)	Ar (%)	CO_2 (ppm)
2030.0	64.65	78.25	20.50	0.95	3500
2030.5	65.30	78.28	20.37	0.95	4000

they found a uniform $\delta^{18}O$ concentration, higher than the mid-Holocene value. They conclude that the ice, therefore, was deposited in a warmer period and perhaps at a lower elevation. From measurements on temperate ice from Alpine regions it is known that the gas content is generally in the order of 30 ml air per kg ice. It is, however, not uncommon to measure gas contents and gas compositions very similar to the ones given in Table 4. Therefore, based on gas measurements, we at least could not exclude the possibility that the ice from the silty layer was temperate at the time of formation.

Acknowledgments. We thank C.C. Langway and W. Dansgaard for excellent collaboration in GISP.

This work has been supported by the Swiss National Science Foundation; the U.S. National Science Foundation, E.P. Todd, F.S.L. Williamson, and R.L. Cameron; the U.S. Department of Energy, R. Dahlmann; and the University of Bern, Switzerland.

We would also like to thank R. Zumbrunn and B. Lehmann for valuable discussions; H. Hofer and D. Rauber for assistance in measurements; and H. Rufli and W. Bernhard for the construction of the crusher.

References

Broecker W.S. and J. van Donk, Isolation Changes, Ice volumes, and the O[18] record in deep-sea cores, *Reviews of Geophysics and Space Physics, 8*(1), 169–197, 1970.

Dansgaard W., H.B. Clausen, N. Gundestrup, C.U. Hammer, S.F. Johnsen, P.M. Kristinsdottir, and N. Reeh, A new Greenland deep ice core, *Science, 218,* 1273–1277, 1982.

Gundestrup N., and S.J. Johnsen, A battery powered, instrumented deep ice core drill for liquid filled holes, this volume, 1985.

Herron, M.M., S.L. Herron, and C.C. Langway Jr., Climatic signal of ice melt features in southern Greenland, *Nature, 293,* 389–391, 1981.

Herron, S. L. and C. C. Langway, Jr., The debris-laden ice at the bottom of the Greenland Ice Sheet, *Journal of Glaciology, 23* (89), 193–207, 1979.

Herron, S., M. Herron, and C.C. Langway, Jr., The effect of seasonal melting of volume of air and crystal size in polar ice cores (draft manuscript), 1982.

Neftel A., H. Oeschger, J. Schwander, B. Stauffer, and R. Zumbrunn, Ice core sample measurements give atmospheric CO_2 content during the past 40,000 yr., *Nature, 295* (5846), 220–223, 1982.

Neftel A., H. Oeschger, J. Schwander, and B. Stauffer, CO_2 concentration in bubbles of natural cold ice, *Journal of Physical Chemistry,* 87 (21) 4116–4120, 1983.

Oeschger H., B. Stauffer, A. Neftel, J. Schwander and R. Zumbrunn, Atmospheric CO_2 content in the past deduced from ice core analyses, *Annals of Glaciology, 3,* 227–232, 1982.

Oeschger H., The contribution of ice core studies to the understanding of environmental processes, this volume, 1985.

Shoji H., and C.C. Langway, air hydrate inclusions in fresh ice core, *Nature, 298* (5874), 548–550, 1982.

Stauffer B., W. Berner, H. Oeschger, and J. Schwander, Atmospheric CO_2 history from ice core studies, *Zeitschrift für Gletscherkunde und Glazialgeologie. 16*(2), 1982.

Weertman J., Comparison between measured and theoretical temperature profiles of the Camp Century, Greenland, borehole, *Journal of Geophysical Research, 73*(8), 2691–2700, 1968.

Zumbrunn R., A. Neftel, and H. Oeschger, CO_2 measurements on 1 cm^3 ice samples with an IR-laserspectrometer (IRLS) combined with a new dry extraction device, *Earth and Planetary Science Letters, 60,* 318–324, 1982.

CONTINUOUS IMPURITY ANALYSIS ALONG THE DYE 3 DEEP CORE

C. U. Hammer, H. B. Clausen, and W. Dansgaard

Geophysical Isotope Laboratory, University of Copenhagen, Denmark

A. Neftel

Physics Institute, University of Bern, Switzerland

P. Kristinsdottir

University of Iceland, Reykjavik, Iceland

E. Johnson[1]

Physical Laboratory II, University of Copenhagen, Denmark

Abstract. Preindustrial Greenland ice-sheet impurities consist of marine, continental, volcanic, stratospheric, and extraterrestial material. In order to estimate the contribution of the various impurity sources, the concentration of insoluble and soluble material was measured on the Dye 3 deep core.

During nonvolcanic periods in the Holocene the oceans and the continents dominate as source regions, but the stratospheric HNO_3 component is probably an important contributor to the generally acid character of the ice. In volcanic periods the continuous acidity profiles reveal several large volcanic eruptions that strongly contribute to the precipitation chemistry up to a few years after the eruptions.

Ice from the Wisconsin glaciation has 3 to 70 times higher dust concentrations than does Holocene ice in both Greenland deep cores. The concentrations of all major impurities, soluble as well as insoluble, are strongly correlated with $\delta(^{18}O)$. Detection of individual volcanic eruptions by acidity measurements is prevented because Wisconsin ice is generally alcalic, and chemical detection is hampered by the high and variable impurity levels. However, Byrd-core analyses show that Antarctica is better suited for this kind of analysis, because the Wisconsin ice is acidic, and the impurity level is much lower than in Greenland.

Introduction

Polar ice-core analysis is closely related to changes in the meteorology and chemical composition of the mid-troposphere over the ice sheets. Such changes are to various degrees reflected in ice cores; the actual mechanism of information transfer from atmosphere to ice being dependent on the parameters under study.

While some information derives directly from the ice *per se*, like the isotopic composition [Dansgaard et al., 1984], another important information group comes from atmospheric trace substances trapped in the ice [Langway, 1970]. Apart from the entrapped air (bubbles) [Stauffer et al., 1984] the ice sheets conserve the above type of information mainly via past accumulation of snowfalls. Dry fallout on the ice sheets, except on marginal zones, contributes only little to bulk impurity concentrations in an ice core; only large particles (>5–$10 \mu m$) of extraterrestrial origin or pollen may show a large percentage of the dry fallout.

From previous studies (for example, Herron [1982], Cragin et al. [1977], Hammer [1977a], Risbo et al. [1981]) we know that the bulk impurity composition of ice cores consists of:

1. Microparticles (insoluble or slightly soluble in water); continentally derived and mainly clay like (Holocene). During the ice age a strong contribution of alkaline, Ca rich, dust was added in the northern hemisphere (proposed origin; ocean free continental shelves).
2. Sea salts; mainly NaCl and some SO_4^{2-}, Ca^{2+}, Mg^{2+}, and K^+;
3. Continentally derived strong electrolytes: NH_4^+ via NH_3, NO_3^-, SO_4^{2-}, H^+ and some Ca^{2+}, Mg^{2+}, K^+, and CO_3^{2-} (or as HCO_3^-);
4. Stratospherically derived strong electrolytes (periods of little volcanic activity: HNO_3);
5 Volcanically produced strong electrolytes: H_2SO_4, HCl, and HF, and salts of these acids; and,
6. Organic material.

Table 1 shows typical composition of Holocene Dye 3 ice. In recent times anthropogenic sources, especially of nitrate and sulphate, may contribute to the Greenland Ice Sheet bulk impurity deposition, but the subject is still controversial.

Considering that the ice sheets consist of precipitations through more than 100,000 years, it is an almost impossible task to perform an exhaustive concentration analysis of all relevant trace substances; even if the analysis is confined to major bulk constituents. As will be described below there is a way out of this problem because for parameters such as acidity (measured on solid ice), electrical conductivity (melted samples), microparticle concentra-

[1] Now at: Geophysical Isotope Laboratory, University of Copenhagen, Denmark.

TABLE 1

Dust		50	μgr./kg ice
Cations	H^+	1.2	μequiv./kg ice
	NH_4^+	0.3	μequiv./kg ice
	Na^+	0.4	μequiv./kg ice
(Mg^{++}, Ca^{++}, K^+)		0.1	μequiv./kg ice
Sum of cations		2.0	μequiv./kg ice
Anions	NO^-	1.0	μequiv./kg ice
	SO_4^-	0.5	μequiv./kg ice
	Cl^-	0.5	μequiv./kg ice
Sum of anions		2.0	μequiv./kg ice

tions (melted samples) and pH (melted samples) it is possible to analyze the core continuously and with a high resolution along the core.

Continuous Measurements

To cover an ice core by continuous analysis makes little sense, if continuity is obtained by keeping a low resolution, for example, 10 samples along a 2 km long ice core. The samples may be representative, but little can be derived from them as most interesting features are averaged out. Annual averages, seasonal changes, individual precipitations, and even long-term changes (more than 100–10,000 years) may be lost. Seasonal changes are of particular interest to glaciology and climatology as they are the basis for well dated time series [Hammer et al., 1978]. Individual precipitations have received less notice partly due to the lack of data and partly due to difficulties in interpreting them.

In order to improve the above situation, some new techniques were introduced and applied in the field during the Dye 3 deep drilling, 1979–81 [Gundestrup et al., 1985]. Selected parameters and techniques of the bulk impurities in ice cores can also be grouped according to their physio-chemical properties:

1. Microparticles;
2. Strong electrolytes;
3. Water soluble organics.

Only the first two groups will be discussed.

1. Microparticles: Generally a large fraction of the microparticle mass is associated with particles having radii in the range $0.1\ \mu m < R < 2 \mu m$, though the upper limit of R during the ice age, especially during periods of high dust concentration, was slightly higher [Petit et al., 1981].

If the amount of dust is inferred from light scattering approximately 90° to a laser beam having $\lambda = 0.6328\ \mu m$ and passing through a melted ice sample, it is possible to estimate the microparticles in the stated size range having refractive indices around 1.10 to 1.25 (relative to water). The resonances in the scattering intensity over the size range are partly blurred due to irregular particle shapes and because all microparticle size distributions can be approximated to a Junge distribution [Junge, 1963] having β values in the range of approximately 2.5 to 3.5, such resonances will have a relatively small impact on the accuracy of the estimated mass per kilogram of ice. Ice age values will probably be somewhat underestimated, but the technique is sufficiently precise for ice-core analysis, as has been shown previously [Hammer, 1977b].

By pumping meltwater from a melted groove along the cleaned ice core surface into the light scattering instrument it was possible to continuously record dust concentrations. The resolution could be varied between 50 to 0.1 cm along the core depending on the desires of the analyst.

2. Strong electrolytes: Among the strong electrolytes the strong acids play an important role in ice-core analysis [Hammer, 1980] (apart from the generally alkaline ice age ice). Due to the high equivalent conductance of the hydrogen ion (H_3O^+) the strong acids contribute strongly to the electrical conductance of polar ice-core meltwater (Holocene). By measuring the meltwater conductivity as well as the acidity of the ice core it is possible to judge the amount of strong electrolytic impurities in the ice.

The acidity can be measured with a resolution of a few millimeters (solid conductivity) [Hammer, 1980] and the electrical conductivity can be measured on the same water, which is used for microparticle measurements. The water used for conductivity measurements was passed on to a pH meter. Even though the acidity of an acid ice core can be obtained by solid conductivity measurements (ECM), this does not apply to alkaline ice. In the latter case pH measurements were used to infer the alkalinity of the ice.

Techniques. Each recovered ice core was split parallel to its central axis into a large and a small fraction. The splitting was accomplished by a special saw built by the University of Bern. The saw was also used to shape and clean the plane surface of the large fraction of the divided ice core with a microtome knife. The core, fixed in a special frame, was brought to a cold cave ($T < -10°C$),

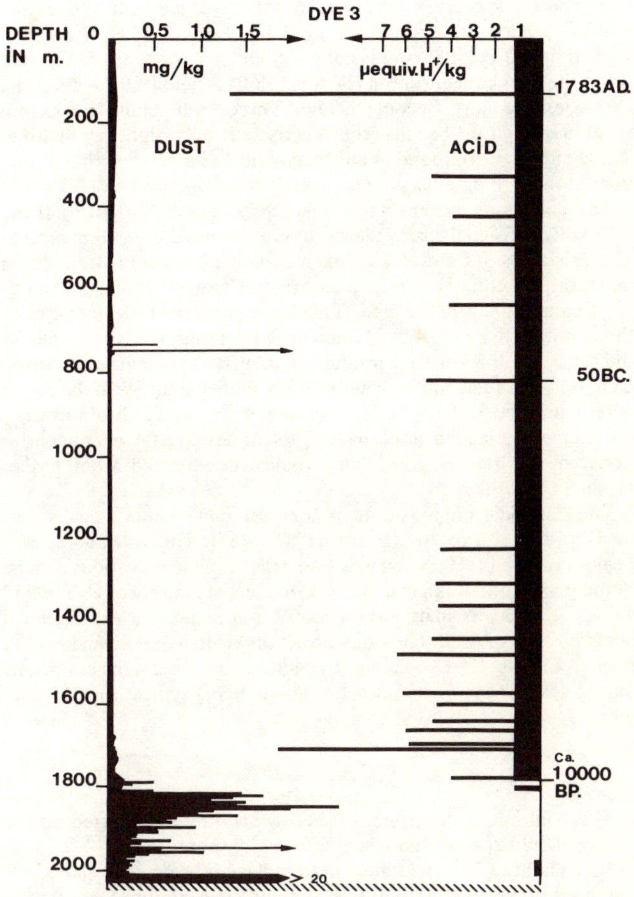

Fig. 1. Dust concentrations and acidity along the 2037 m long deep ice core from Dye 3, South Greenland, 65° 11′N, 43° 49′ W. The high acidity peaks are annual average values exceeding approximately 3.5 μ equivalent H^+/kg of ice; probably all due to major past volcanic eruptions.

where the acidity measurements were performed [Neftel et al., 1985].

The smaller part of the core was sampled for δ (^{18}O) and then the remaining part of this core fraction was cleaned by a microtome knife (the plane surface). A solid-gold tip, replacing a standard soldering tip in a temperature regulated soldering instrument, was used to melt a groove along the core. The meltwater was pumped into a light-scattering instrument, after removal of air bubbles, and passed on to a flow-type conductivity cell. Finally the water entered a pH meter, with a digital output reading. The pH data were recorded directly on the corresponding dust and conductivity output curves. The whole setup had to be heated and the temperature in the flow system was maintained at +30°C, while ambient temperature varied in the range of −2 to −5°C.

Results

Acidity (ECM)

The acidity was measured on the entire Dye 3 core. Most data still needs to be digitized but one paper on acid traces in the core from Icelandic eruptions has been published [Hammer, 1984]. In Figure 1 (right) some of the biggest volcanic acid peaks are indicated (one year averages and larger than approximately 3.5 μ equivalent H^+/kg ice). The thick black strip represents the avergae acidity over the Holocene period, i.e. approximately 1 μ equivalent H^+/kg. From 1786 m to bedrock the core is generally alkaline; only in a few cases, when the dust concentration (Figure 1, left) is below 200–400 μgr/kg ice, does the core become acidic. Three high-acidity peaks are omitted in Figure 1 because they are associated with heavy melting. These signals were found in core segments being filled with percolations (for more details, see Hammer [1984]).

The core is at present dated absolutely back to 200 BC and the high volcanic signals from 200 BC to present have been compared to similar signals in Central and North Greenland ice cores from Crete and Camp Century [Hammer et al., 1980; Hammer, 1983a]. Generally volcanic signals in the Dye 3 core (μ equivalent H^+/kg ice) are 1.7 times lower than in Central Greenland. This ratio is almost equal to the ratio of 1.8 for fission products (^{90}Sr) from nuclear bomb tests. Generally the Icelandic volcanoes show higher acidities in the Dye 3 core. Contrary to the Icelandic volcanoes there are indications that, for example, Alaskan and Japanese volcanoes give lower concentrations in the Dye 3 core than would be expected from fission products.

The chemical composition of some of these acidity peaks was measured in situ by M. Herron [1982; 1985]. The volcanic acidity peaks consist of H_2SO_4, HCl, and HF, though in various ratios. Some peaks, like Lakigigar (1783 AD), consist entirely of H_2SO_4 and sulphuric acid was an important contributor in most cases. Chemical analysis also showed that a large fraction of the nonvolcanic acidity background is HNO_3. Generally nitrate and acid concentrations varied seasonally; higher concentrations being typical for the summer.

Dust

The continuous recording of dust concentrations started during the 1980 field season and had the following purposes:

1. To aid the δ(^{18}O) dating and to replace it when seasonal δ(^{18}O) becomes obliterated by diffusion in the deeper strata of the core;
2. To investigate in detail the expected high dust concentrations during the ice age part of the core.

The high resolution of the method also enables individual precipitations to be seen some 5,000 years back in time (results will be published elsewhere).

The overall view of the dust profile along the core can be seen in Figure 1 (left). Part of the Holocene profile (down to 1786 meter-accumulated core length) is incomplete, but the difference between the Holocene and the Wisconsin ice age is clearly seen. The bottom 25 m are discolored silty ice (granite), which contains small pebbles in the lower part. In this segment of the core one must expect the original stratification to be lost.

The two arrows indicate *visible* dust layers, but of different origin. The upper layer at 751.51 m consists of a few centimeters of yellow to brown banded layering. The dust layer cannot come from the bedrock, as this would require very high mountains within some 5 to 10 km upstream from the drill site. Though the bedrock topography in the Dye 3 region is quite irregular [Overgaard and Gundestrup, 1985] the hills and mountains are too small to disturb the stratification at the depth in question. The dust concentration is high (70 times the Holocene average) over a whole year, i.e. from summer 174 A.D. to summer 175 A.D. as corresponding to the depth interval 751.20 to 751.51 m.

The particle-size distribution of the dust indicates an atmospheric transportation mechanism. There is nothing unusual to be seen in the seasonally varying $\delta^{18}O$ profile, but the ice is slightly alkaline, which is unique for the Holocene ice, at least in the accumulation of an entire year. Though the dust itself is not very alkaline a high atmospheric dustload must have preceded the deposition on the ice sheet, as the dust apparently was sufficient to neutralize the atmospheric background acids. The layer is an anomaly. It does not seem to be volcanic ash, as nearly all particles are crystalline or microcrystalline. For a more thorough discussion of the layer see Hammer [1984].

The other visible layering is situated approximately 88 meters above the bedrock. There are several indications that it derives from the bedrock (see below) and therefore indicates that the continuity of the annual layering is broken or partly broken below 88 m.

Ice Age Part of the Core

In Figure 2 δ O^{18}, dust, and alkalinity of the ice are shown (from left to right). At 1786 m depth is the transition from ice ae to Holocene. The alkalinity could only be estimated from 1860 m and downwards because the pH probe broke during core processing between 1750 to 1860 m depth.

All three data sets are given as 55 cm average samples. Actually $\delta^{18}O$ were measured as 5 and 2.5 cm samples, dust continous (5 cm resolution at this depth) and alkalinity (pH) continuous (10 cm resolution). The $\delta^{18}O$ is given in the δ scale, dust in mg/kg ice and alkalinity in μmho · cm^{-1}. The latter unit has the advantage that it can be easily compared to the meltwater's total electric conductance, which is also given in μmho · cm^{-1}. The alkalinity is calculated by using the pH measurements as a starting point. The pH is measured as a last step in the flow system and most carbonate particles are dissolved when the pH meter is read. The flow system has been calibrated for ambient CO_2 uptake, which depends on the barometric pressure (Dye 3 is at 24% m asl) and the temperature in the flow system. The design of the system also influences the CO_2 uptake. It is assumed that the alkalinity stems from $CaCO_3$ microparticles (see, for example, Cragin et al. [1977]). This is of course an oversimplification, but as the pH ranges from 5.7 to 6.4 for alkaline ice (atmospheric CO_2 present in the samples) most alkaline material will be present as $Ca(HCO_3)_2$. When the dust concentrations are very high, i.e. exceeding approximately 2 mg/kg ice, it is difficult to obtain chemical equilibrium with the present setup, but the alkalinity, as defined here, does show a remarkable correlation with the dust concentration (Figure 2).

The thin curve over the alkalinity is the total electrical conductance of the meltwater. The difference between the thin curve and

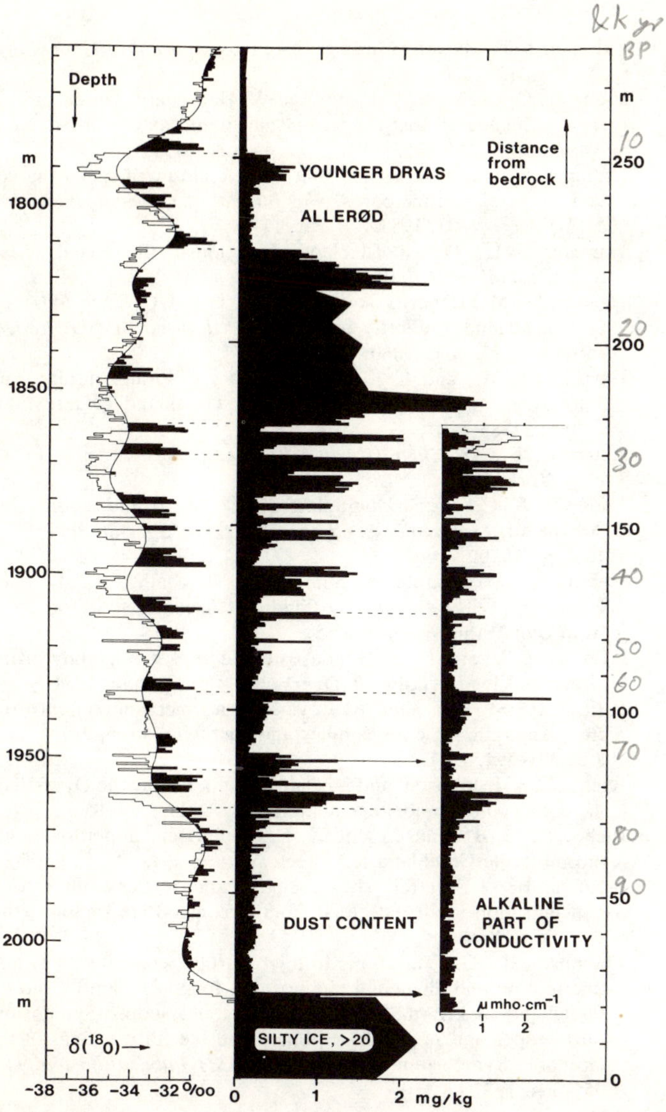

Fig. 2. Ice age part of the Dye 3 deep core showing $\delta^{18}O$, dust concentrations, and alkalinity in 55 cm average samples. From 1821 to 1852 m only selected segments were measured for dust, which gives the dust profile a sawtooth appearance in this depth region.

the alkalinity is due to sea salt, nitrates, and volcanic impurities (occasionally). Note that high dust concentrations, high alkalinity, and high sea-salt concentrations, etc. vary in a parallel manner. The $\delta^{18}O$ profile shows that low δs correspond to high impurity content. The discussion of the $\delta^{18}O$ profile is treated elsewhere [Dansgaard et al., 1982; 1985].

Note that below 88 m above bedrock (visible dust layer), the dust concentration and alkalinity do not correlate well. Also the very high amounts of dust in the silty ice have little effect on the alkalinity. The bedrock dust is granitic and the visible dust layer seems also to be granitic (from preliminary analysis of the dust). Hence, it seems that the 88 m visible layer has been introduced into the ice when the ice flow was inhibited by a large mountain upstream from the Dye 3 drill site. The flow of the ice is treated elsewhere [Reeh et al., 1985].

The dating of the ice age ice is difficult, as the annual layers are very small; varying from 2 cm in the Allerod period to less than 1 mm close to bedrock. It is possible to improve the resolution of the dust measurements to perhaps 10 to 100 μm, and thus resolve the annual variations of the dust concentrations.

It is a prerequisite for such a dating that the amount of precipitation during the ice age be sufficiently high. Insufficient precipitation combined with strong winds would tend to obliterate seasonal variations even though microparticles hardly diffuse in the ice core. Dust measurements on the Camp Century core indicate that the precipitation during the ice age may have been a factor of approximately 2 less than today [Hammer et al., 1978]. If this holds for the Dye 3 region it should be possible to date the Dye 3 core stratigraphically; at least down to the 88 m visible layer.

More detailed profiles exist for the rapid changes in δ, dust and alkalinity which can be seen in Figure 2. These changes and their cause are highly interesting, however the data treatment is not yet finished. These changes are also seen in the Camp Century core and have been discussed by Dansgaard et al. [1985].

Comparison with the Antarctic Byrd Core

Impurity analyses of the 2184 m long Byrd core [Cragin et al., 1977] show such low impurity content in the ice age part of the core that atmospheric acid impurities over Antarctica should not be neutralized during the ice age. This was recently confirmed by continuous acidity measurement on the Byrd core [Hammer, 1983b; Hammer and Clausen, unpublished data]. The acidity profile proved to vary seasonally even during the ice age, and several major volcanic eruptions were detected. Figure 3 shows an example of acid fallout from a large volcanic eruption.

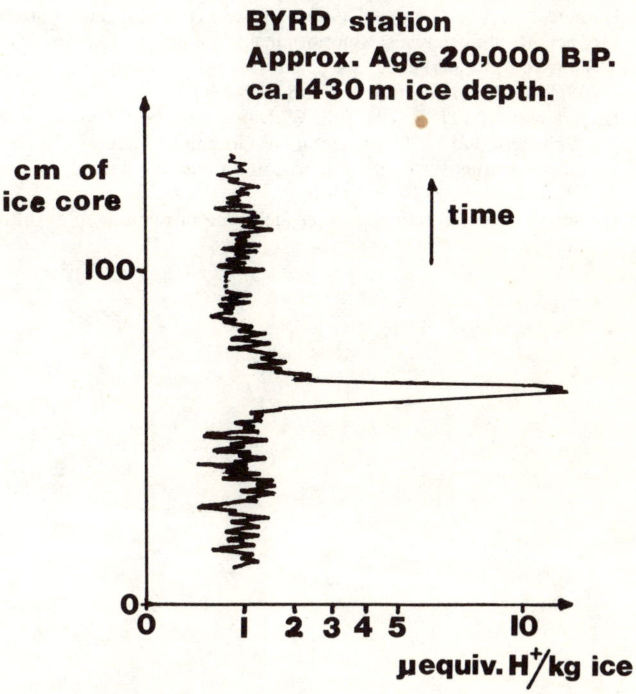

Fig. 3. Acidity profile over 1.25 m of the Byrd core. Approximate age 20,000 BP, depth 1430 m. Acid fallout from a major eruption can be seen. The annual layer thickness is approximately 2 cm at this depth.

The rapid changes in $\delta^{18}O$ as seen in Figure 2 seem to be mainly a northern hemisphere phenomenon as well as are the high amounts of alkaline dust. The high dust concentrations in Greenland ice age ice [Thompson, 1977] are most likely caused by the large source regions for loess, while the alkaline source could be the continental shelves in combination with a lower sea level during the ice age. Particularly the ocean bottom north of the USSR may have been a large source region.

The ice sheets of Greenland and Antarctica are both important in order to get a more comprehensive outlook on past climate and past atmospheric impurity composition.

References

Cragin, J. H., M. M. Herron, C. C. Langway Jr., and G. A. Klouda, Interhemispheric comparison of changes in the composition of atmospheric precipitation during the Late Cenozoic Era, in *Polar Oceans*, edited by M. J. Dunbar, pp. 617–631, Arctic Institute of North America, Calgary, Alberta, 1977.

Dansgaard, W., H. B. Clausen, N. Gundestrup, C. U. Hammer, S. J. Johnsen, P. Kristinsdottir, and N. Reeh, A new Greenland deep ice core. *Science, 218*, 1273–1277, 1982.

Dansgaard, W., H. B. Clausen, N. Gundestrup, S. J. Johnsen, and C. Rygner, Dating and climatic interpretation of two deep Greenland ice cores, this volume, 1985.

Gundestrup, N. S., S. J. Johnsen, and N. Reeh, ISTUK a deep ice core drill system, Second International Symposium on Ice Drilling Technology, August 30–September 1, 1982, Calgary, Canada, *Special Report*, U.S. Army Corps of Engineers, Cold Regions Research and Engineering Laboratory, Hanover, N. H., in press, 1985.

Hammer, C. U., Past volcanism revealed by Greenland Ice Sheet impurities, *Nature, 270*, 482–486, 1977a.

Hammer, C. U., Dating of Greenland ice cores by microparticle concentration analyses, Symposium on Isotopes and Impurities in Snow and Ice (Proceedings of the Grenoble Symposium, 1975), *IAHS-AISH Publication No. 118*, 297–230, 1977b.

Hammer, C. U., H. B. Clausen, W. Dansgaard, N. Gundestrup, S. J. Johnsen, and N. Reeh, Dating of Greenland ice cores by flow models, isotopes, volcanic debris, and continental dust. *Journal of Glaciology, 20*, 3–26, 1978.

Hammer, C. U., Acidity of polar ice cores in relation to absolute dating, past volcanism, and radioechoes, *Journal of Glaciology, 25*, 359–372, 1980.

Hammer, C. U., H. B. Clausen, and W. Dansgaard, Greenland ice sheet evidence of post-glacial volcanism and its climatic impact, *Nature, 288*, 230–235, 1980.

Hammer, C. U., Initial DC current in the build-up of space charges and the acidity of ice cores. *The Journal of Physical Chemistry, 87*(21), 4099–4103, 1983.

Hammer, C. U., Traces of Icelandic eruptions in the Greenland Ice Sheet, *Jökull*, 51–65, 1984.

Herron, M. M., Impurity sources of F^-, Cl^-, NO_3^- and SO_4^{2-} in Greenland and Antarctic precipitation. *Journal of Geophysical Research, 87,* 3052–3060, 1982.

Herron, M. M., and C. C. Langway Jr., Chloride, nitrate, and sulfate in the Dye 3 and Camp Century, Greenland ice cores, this volume, 1985.

Junge, C. E., *Air Chemistry and Radioactivity*, Academic Press, New York, 1963.

Langway, C. C., Jr., Stratographic analysis of a deep ice core from Greenland, *Geological Society of America, Special Paper 125*, 186 pp., 1970.

Neftel, A., M. Andrée, J. Schwander, B. Stauffer, and C. U. Hammer, Measurements of a kind of DC-conductivity on cores from Dye 3, this volume, 1985.

Overgaard, S., and N. S. Gundestrup, Bedrock topography of the Greenland Ice Sheet in the Dye 3 area, this volume, 1984.

Petit, J. R., M. Briat, and A. Royer, Ice age aerosol content from East Antarctic ice core samples and past wind strength. *Nature, 293*, 391–394, 1981.

Reeh, N., S. J. Johnsen, and D. Dahl-Jensen, Dating the Dye 3 deep ice core by flow model calculations, this volume, 1985.

Risbo, T., H. B. Clausen, and K. L. Rasmussen, Supernovae and nitrate in the Greenland Ice Sheet, *Nature, 294*, 637–639, 1981.

Stauffer, B., A. Neftel, H. Oeschger, and J. Schwander, CO_2 concentration in air extracted from Greenland ice samples, this volume, 1985.

Thompson, L. G., Variations in microparticle concentration, size distribution and elemental composition found in Camp Century, Greenland, and Byrd Station, Antarctica, deep cores. Symposium on Isotopes and Impurities in Snow and Ice (Proceedings of the Grenoble Symposium, 1975), *IAHS-AISH Publication No. 118*, 351–364, 1977.

THE HISTORICAL RECORD OF ARTIFICIAL RADIOACTIVE FALLOUT FROM THE ATMOSPHERE IN POLAR GLACIERS

Minoru Koide and Edward D. Goldberg

Scripps Institution of Oceanography, La Jolla, California

Abstract. Polar glaciers maintain a record of artificial radionuclide atmospheric fallout which can be developed on a year to year basis. Some homogenization of the record results from summer melting and subsequent melt-water percolation. Fallout from the U.S. dominated tests of the 1950s differs from that of the U.S.S.R. dominated tests in the 1960s. Further, differences in the atmospheric transport of the transuranics, alkalis, alkaline earths, and tritium were observed.

Introduction

Polar ice sheets maintain a record of the changes in atmospheric-aerosol contents over time and are especially useful in studies of the alterations due to the activities of human society. We have developed the historical record of artificial radioactive fallout primarily from weapons tests over the past quarter of a century, and herein we review and assess our findings.

Glacial sediments are especially attractive in such studies for a number of reasons. First of all, a number of geochronologies, both radiometric and stratigraphic, allow the introduction of time frames into the glacial column. Our work which utilizes primarily Pb-210 for determining deposition rates and, to a lesser extent, relies on the identification of uniquely dated episodes that define horizons in the fallout pattern indicates that ages of the glacial strata can usually be determined to within plus or minus a year over the last three decades. For example, the average accumulation rate based upon a Pb-210 profile for the Greenland ice sheet at Dye 3 is 52 g $H_2O/cm^2/yr$ which may be compared with the stable oxygen isotopic analysis which gives a value of 49 $g/cm^2/yr$ [Koide et al., 1982; Reeh et al., 1978].

A second advantage in using glaciers to investigate atmospheric compositions over time is their widespread occurrences about the surface of the earth. Since the principal wind systems generally move zonally, the long range (thousands of kilometers) transport of aerosols may well be recorded in glaciers situated at nearly the same latitudinal band as the source. However, the polar glaciers, in contrast to their alpine counterparts, maintain a more reliable stratigraphic record. In the Arctic and Antarctic ice sheets there generally is a smaller amount of summer melting with a consequential smaller smear in the record from percolation. The principal disadvantage in the employment of glaciers is this melting. Approximately 8% of each year's snowfall at Dye 3 was melted and refrozen as ice layers [Herron et al., 1981]. The downward percolating water can enter strata deposited several years previously. Still, we shall point out subsequently that perhaps some of the anomalies in the glacial record may well be explained by this small amount of melting. Thus the potential use of alpine glaciers as historical records may be limited to those in which the melting is of the same order of magnitude as that of the polar ice sheets.

We emphasize the necessity to study groups of artificial radionuclides to understand both the meteorology of transport and their atmospheric chemistries. Since the times and intensities of weapons testing are moderately well documented, the glacial record can best be interpreted where there is an excess of information or redundancy in the data to compensate partially for any alteration of the record due to percolation.

Source Functions

There are several distinct periods of atmospheric nuclear weapons testing based upon the activities of one or more countries. Large yield events initially were carried out during 1952–1954 by the U.S. with the Ivy-Mike and Castle-Bravo thermonuclear activities. Up to the moratorium on such tests from November 1958 to September 1961, the U.S. dominated in the introduction of materials to the

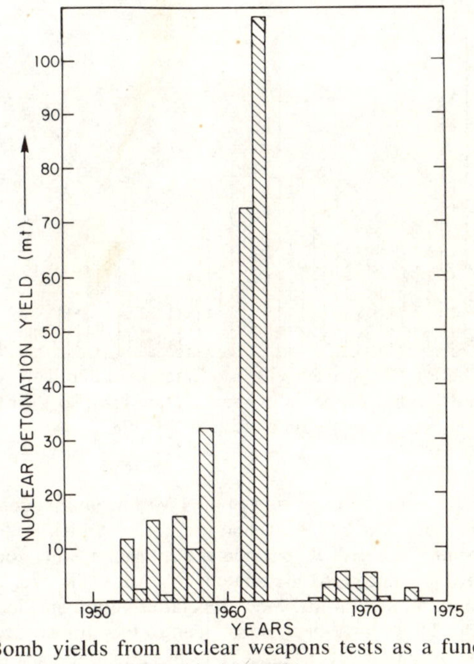

Fig. 1. Bomb yields from nuclear weapons tests as a function of time. (From Carter and Moghissi [1977].)

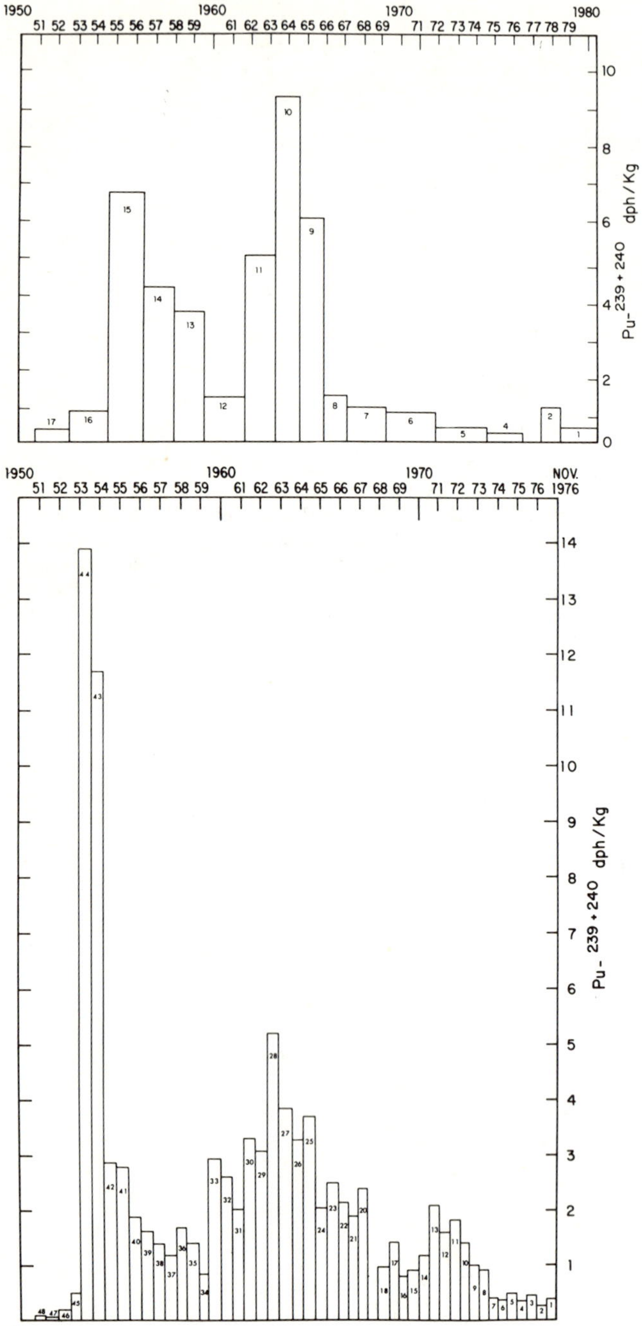

Fig. 2. The Pu-239+240 activity as a function of depth and age in the Greenland (Dye 3)(upper plot) and Antarctica (Ross Shelf)(lower plot) ice sheets. (From Koide et al. [1979; 1982].)

atmosphere with the explosion of 48.2 Mt (megatons) compared to that of the U.S.S.R. (19.9 Mt) and the U.K. (18 Mt). Following the moratorium the U.S.S.R. became the primary testing country and between 1961 and 1962 detonated 168 Mt while the U.S. was involved with but 12.4 Mt. Since this latter period smaller amounts of nuclear debris have been introduced to the atmosphere by tests from the Republic of France and the Republic of China. The amounts of radioactivity introduced to the atmosphere from these tests are given in Figure 1 taken from the data of Carter and Moghissi [1977].

The radionuclides enter the atmosphere in the gaseous or particulate phases and are produced as fission products, unburned transur-

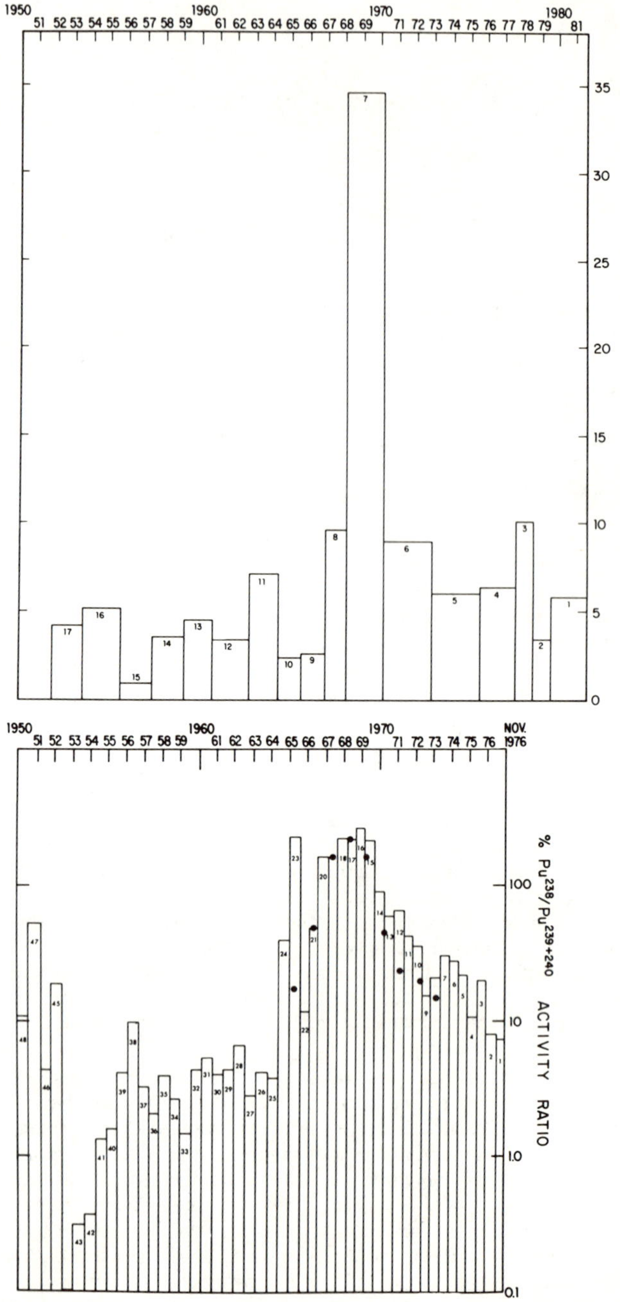

Fig. 3. The Pu-238/Pu-239+240 activity ratio as a function of depth and age in the Greenland (Dye 3)(upper plot) and Antarctica (Ross Shelf)(lower plot) ice sheets. The heavy circles on the Ross Shelf plot are unpublished data for plutonium in atmospheric particulates measured by the Environmental Measurements Laboratory of the U.S. Department of Energy. The sample 44 has a value of slightly greater than 0.1. (From Koide et al. [1979; 1982].)

TABLE 1. Inventories of Artificial Radionuclides in Polar Ice Sheets [Koide et al., 1982]

	$^{239+240}Pu$		^{137}Cs		^{90}Sr		^{3}H*		$^{239+240}Pu$ / ^{137}Cs	$^{239+240}Pu$ / ^{90}Sr	^{137}Cs / ^{90}Sr
	Inventory	(%)	Inventory	(%)	Inventory	(%)	Inventory	(%)			
ARCTIC											
Dye 3 Greenland (65°11′N, 43°50′W)	0.3		21.3		16.7		2.27×10^{-3}				
Postmoratorium		59		65		72		87	0.012	0.013	1.1
Premoratorium		41		35		28		13	0.016	0.023	1.6
South Dome, Greenland (63°31.6′N, 44°34.5′W)	0.7										
Postmoratorium		56									
Premoratorium		44									
ANTARCTIC											
J-9 (82°22′S, 168°40′W)	0.04		2.8		2.7		1.12×10^{-4}				
Postmoratorium		57		68		77		85	0.012	0.011	0.9
Premoratorium		43		32		23		15	0.019	0.027	1.5
Dome C(74°39′S, 123°10′E)	0.04										
Postmoratorium		36									
Premoratorium		64									

The inventories calculated for all nuclides except $^{239+240}Pu$ were corrected for decay to the time of deposition as derived from ^{210}Pb chronology and are in mCi km^{-2}.
*^{3}H inventory units in g km^{-2}.

anics or induced activities from the hardware of the devices. The physical state of any given nuclide will be an important factor in its travels through the atmosphere and its subsequent deposition on the earth's surface.

Various sites of testing used by different nations were located at markedly different latitudes. The U.S.S.R. tests took place at high latitudinal sites in the northern hemisphere (Novaya Zemlya, 75°N, 55°E; Siberia, 52°N, 78°E and Semipalatinsk, 50°N, 80°E). The

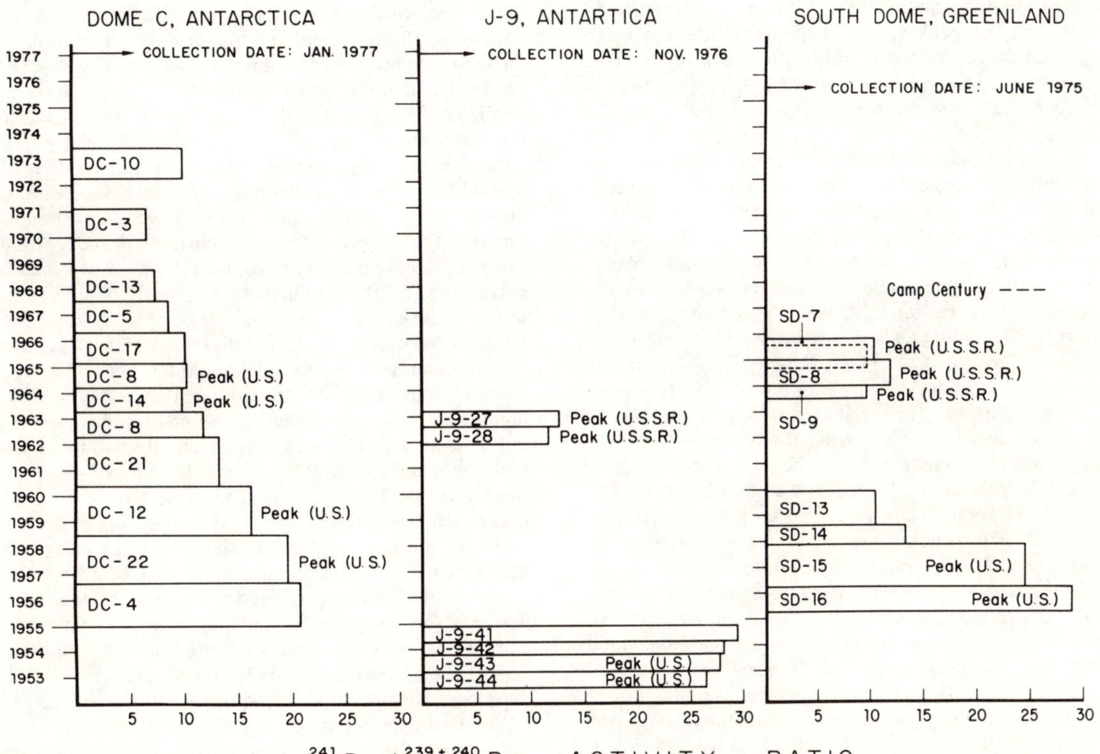

Fig. 4. The Pu-241/Pu-239+240 activity ratios corrected to the time of deposition of polar glaciers plotted against time. Ages are derived from their respective Pb-210 chronologies. The Pu-241/Pu-239+240 ratios are calculated to the time of deposition based on these chronologies. (From Koide and Goldberg [1981].)

TABLE 2. Am-241/Pu-239+240 Ratios in Seawaters [Koide and Goldberg, 1981]

Sample	Collection Date	Location Site	Depth (m)	Am-241/Pu-239+240
Atlantic Ocean	1972–1974		<1000	0.22 (av.)
			>1000	0.31 (av.)
Pacific Ocean	1972–1974	~31°N, 148°W	<1000	0.21 (av.)
			>1000	0.29 (av.)
Mediterranean Sea	1976	43°32′N, 07°32′E	<1000	0.08 (av.)
			>1000	0.28 (av.)
	1975		<1000	0.08
			>1000	0.19

larger U.S. detonations took place at the Pacific Proving Grounds (Eniwetok and Bikini at 11°N, 162°E and 11°N, 165°E respectively and Johnston Island, 11°N, 165°E). The French tests occurred at Mururoa and Fangataufa on the Tuamotu Archipelago (21°S, 137°W) while the Chinese tests were performed at Lop Nor, Sinkiang Province (40°N, 90°E). The French tests may have contributed substantially to the Antarctic ice sheets as a consequence of their proximity.

These spatial and temporal characteristics of the source functions are most important in the interpretation of the glacial records. For example, the early test period, dominated by U.S. shots at low latitudes, is evident in the high Pu-239+240 concentrations in the Antarctic glacial strata deposited between 1953 and 1954 (Figure 2). They are higher than the peak values attributed to the extensive U.S.S.R. tests whose record is found in the strata deposited in the early 1960s. A reversal of the intensities of these two maxima is found in the Greenland ice sheet, clearly closer to the site of the U.S.S.R. detonations.

Radioactive debris introduced to the stratosphere from weapons tests, remains there with a generally accepted residence time of somewhat under a year before returning to the troposphere and then to the earth's surface. During this period extensive mixing can take place between the northern and southern parts of the stratosphere. The mixing is nearly complete and this can be seen from the plutonium inventories (Figure 2 and Table 1) in polar ice sheets. On the basis of atmospheric bomb yields the ratio of post to premoratorium testing is around 70:30 [Carter and Moghissi, 1977]. On the basis of the integrated inventories for plutonium the contributions from premoratorium tests vary between 41 and 64% in both Arctic and Antarctic ice sheets. The values from three of the four sites clusters about 43%. The higher value comes from the work of another laboratory [Cutter et al., 1979]. On the other hand, the Cs-137 and Sr-90 results (Table 1) are in accord with the 70:30 ratio from the bomb yields. This is not unexpected since Cs-137 and Sr-90 are fission products whereas Pu is not. Pu production during weapons testing does not necessarily correlate with total bomb yield, as it is also dependent upon the type of weapon construction.

The burn up of plutonium metal parts from a nuclear-powered satellite (SNAP-9A) that aborted on April 21, 1964, introduced 17 kcuries of Pu-238 to the earth's outer environment. The reentry of the debris from the southern hemisphere stratosphere over the Indian Ocean and then to the earth's surface introduced, in principle, a time horizon to glaciers. Clearly, the mode of injection and the subsequent transport of radionuclides differ from those of weapons-test radionuclides. The ratios of Pu-238 to other artificially produced radionuclides in glacial strata did show changes after 1964.

Characteristic Nuclide Compositions during the Testing Periods

The testing period dominated by the U.S. activity in the early 1950s produced low Pu-238/Pu-239+240 ratios (and very high Pu-241/Pu239+240 ratios, see below) compared with those resulting from tests dominated by the U.S.S.R. activity in the early 1960s (Figure 3). The low ratios were also found in coastal marine sediments deposited at the same time off the coast of Peru [Koide and Goldberg, 1982]. These results indicate that there are characteristic nuclide compositions of fallout produced by weapons testing by different countries.

The SNAP-9A event becomes evident in the 1964 strata (Sample 24) of the Ross Ice Shelf in Antarctica (Figure 3). The burn up altitude of 50 km provided a higher atmospheric injection site than those for the previous weapons test debris. The Pu-238 levels are an order of magnitude greater than those in ice strata preceding the Sample 24. The falloff in the Pu-238 occurs with a half time of longer than a year, greater than the residence time of debris in the stratosphere. This probably reflects a mobilization of the Pu-238 within the ice sheet as a consequence of percolation. Or perhaps it may reflect a transport time of several years for the Pu-238 to reach the lower stratosphere from the burn up altitude. Another explanation is that there is an entry to the glaciers of continental soil debris, containing Pu-238, mobilized by the winds. Finally, one should note in Figure 3 the good agreement in our glacial values of the Pu-238/Pu-239+240 ratios with those of H. L. Volchok (personal communication) taken from air samples in Antarctica.

Pu-241, another nuclide produced in weapons testing, has the highest activity among the plutonium isotopes in atmospheric fallout. It has a half life of 14.4 years and decays to a long lived alpha-emitting daughter, Am-241, with a half life of 433 years. Similar to the Pu-238/Pu-239+240 ratios, the Pu-241/Pu-239+240 ratios in the glacial strata are characteristic of the testing periods. The average value of 26 was found for the fallout from the early U.S. dominated series in the 1950s as compared to that of 12 for the subsequent U.S.S.R. tests when corrected to the time of deposition (Figure 4). There is some support for these differences from previous studies. The U.S. tests Bravo and Mike produced atmospheric debris with Pu-241/Pu-239+240 ratios of 26 and 27 respectively [Diamond et al., 1960, and Hisamatsu and Sakanoue, 1978].

Similarly, the ingrowth of the Am-241 into the strata containing debris from the early U.S. dominated test period is readily seen in the glacial strata [Koide and Goldberg, 1981]. The presence of Am-241 is attributed to the decay of Pu-241 from nuclear testing and the assumption is made that a negligible amount of the parent nuclide was produced during the detonation.

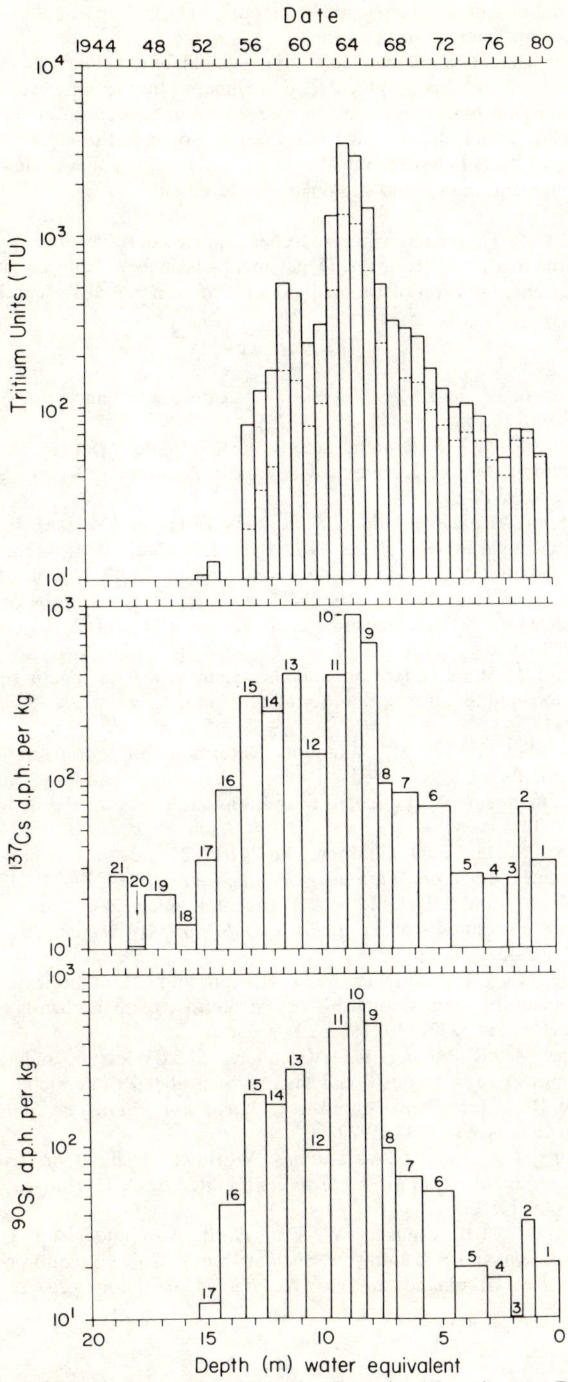

Fig. 5. H-3, Cs-137 and Sr-90 in the Greenland Ice sheet at Dye 3 station as a function of depth.

We have observed that the particulates in older seawater samples (i.e., those waters at great depths) have Am-241/Pu-239+240 ratios higher than those samples from shallower depths (i.e., younger samples) (Table 2). Perhaps some sense of the age of the particulates in these waters can eventually be surmised from the Pu-241/Pu-239+240 ratios.

There are also characteristic Pu-239+240/Cs-137 and Pu-239+240/Sr-90 ratios produced during the two testing periods on the bases of the glacial record (Table 1). High values of the ratios from the U.S. tests (premoratorium period) are found both in the Antarctic and Greenland samples. They are about twice those of the years previous and subsequent to the two weapons test periods. The lower values of these ratios agree well with those found in air samples from Washington, New York and the U.K. taken during the postmoratorium period. No air data are available for comparison with the premoratorium values [Perkins and Thomas, 1980]. Also, the Cs/Sr ratio during the U.S. test period is significantly higher than during the postmoratorium period. Again, there is compelling evidence that weapons debris introduced to the stratosphere become mixed in times that are short with respect to the residence times of particles of about one year on the basis of these characteristic ratios found in both polar ice sheets.

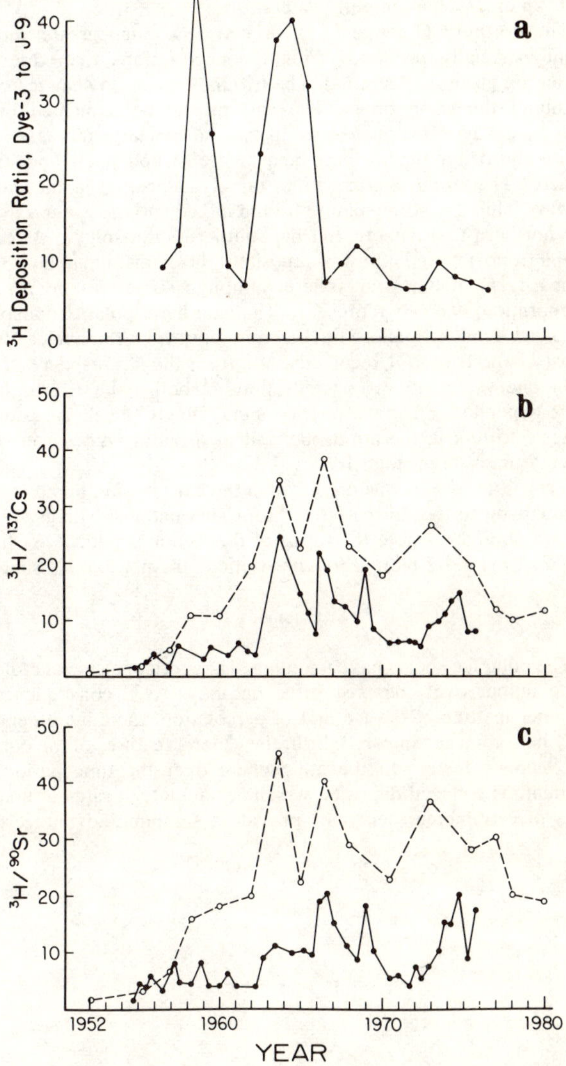

Fig. 6. (a) Ratio of H-3 deposition at Dye 3 to Antarctic Site J-9 as a function of time; (b) H-3/Cs-137 ratio as a function of time; (c) H-3/Sr-90 as a function of time. O, Dye 3; ●, J-9.

Differences in Atmospheric Behaviors of Artificial Radionuclides

In both the northern and southern polar ice sheets the recorded times of arrival of radioactive debris from weapons tests follow a pattern where the transuranics are the first to accumulate, followed by the alkali and the alkaline earth nuclides and then the tritium [Koide et al., 1979; 1981]. The 1955 strata (Level 15, Figures 2 and 5, and Table 1) of the Dye 3 Greenland ice sheet contained contents of artificial radionuclides relative to the total premoratorium period (Levels 16, 15, 14, and 13) in the following ratios: Pu-239+240, 46, Cs-137 or Sr-90, 31 and H-3, 17. Those U.S. test-produced nuclides that moved slower in the atmosphere than Pu-239+240 accumulated in the ice sheet in the years subsequent to 1955. We had previously suggested that the transuranics, in part occurring as particulates in the atmosphere, have the shortest residence times in the stratosphere, while tritium in the gaseous form has the longest. Cs-137 and Sr-90, on such a basis, would have intermediate values for persistence in the stratosphere. This may relate to their hydrophilic natures as ions and the stratospheric times involved in their abilities to form an association with waters.

The northern Greenland glacier shows a much greater impact from artificially produced tritium concentrations than does the Antarctic glacier (Figure 6a). The tritium deposition changes occur at about the same times. The pattern can be explained in the following way. Most nuclear testing produced a large increase in the tritium burden in the northern hemisphere's troposphere and stratosphere. The tropospheric tritium rapidly deposited at the earth's surface while the stratospheric tritium mixed into the water vapor of the northern troposphere and the southern stratosphere. After the initial tropospheric input was deposited, the deposition at Dye 3 was primarily from the stratosphere, although some tritium from reevaporation was also present. The southern polar troposphere received a much smaller initial tritium input from the nuclear tests. Most of the tritium it received came from the stratosphere. Thus, after one year, tritium concentrations in both polar tropospheres were dependent on stratospheric inputs. Due to the short residence time of tritium in the atmosphere, deposition at Dye 3 was about twenty times larger than J-9.

Very little tritium was produced in the early fission bomb tests in comparison to that in the latter fusion detonations. This is seen in Figure 6b and c where the ratios of the fission products Sr-90 and Cs-137 to H-3 are plotted for the Arctic and Antarctic ice sheets.

Overview

The polar ice sheets have maintained a record of artificial radionuclide fallout over the past three decades. As a consequence of summer melting of the ice and of percolation, the annual deposits may be somewhat smeared. Still, the general features of the outputs of weapons testing to the atmosphere over the time period are evident. The glacial deposits, which accumulate at rates of units to tens of centimeters per year, provide a sampling advantage over coastal marine and lacustrine sediments that build up at millimeters to a centimeter or so per year.

The use of polar glaciers in studying global fallout clearly gives the results of two geographical end members. In the past we have emphasized the urgent need to carry out similar measurements on mid-latitudinal glaciers. We suspect that some of these ice sheets may hold a detailed record of artificial radioactivity fallout that will supplement and extend the polar measurements.

Acknowledgments. The researches reported herein were supported under a grant from the Department of Energy, Environmental Programs, Division of Biomedical and Environmental Research.

References

Carter, M.W., and A.A. Moghissi. Three decades of nuclear testing, *Health Physics, 33,* 55–71, 1977.

Cutter, G.A., K.S. Bruland, and R.W. Risebrough. Deposition and accumulation of plutonium isotopes in Antarctica, *Nature, 279,* 628–629, 1979.

Diamond, H., P.R. Fields, C.S. Stevens, M.H. Studier, S.M. Fried, M.G. Inghram, P.C. Hess, G.L. Pyle, J.F. Mech, W.R. Manning, A. Ghiorso, S.G. Thompson, G.H. Higgins, G.T. Seaborg, C.I. Browne, H.L. Smith, and R.W. Spence, Heavy isotope abundances in "Mike" thermonuclear device, *Physical Review, 119,* 2000, 1960.

Herron, M.M., S.L. Herron, and C.C. Langway, Jr., Climatic signal of ice melt features in Southern Greenland, *Nature, 293,* 389–391, 1981.

Hisamatsu, S., and M. Sakanoue, Determination of transuranium elements in a so-called "Bikini Ash" sample and in marine sediment samples collected near Bikini atoll, *Health Physics, 33,* 301–307, 1978.

Koide, M., and E.D. Goldberg, Pu-241/Pu-239+240 ratios in polar glaciers, *Earth and Planetary Science Letters, 54,* 239–247, 1981.

Koide, M., and E.D. Goldberg, Transuranic nuclides in two coastal marine sediments off Peru, *Earth and Planetary Science Letters, 57,* 263–277, 1982.

Koide, M., E.D. Goldberg, M.M. Herron, and C.C. Langway, Jr., Transuranic depositional history in South Greenland firn layers, *Nature, 296,* 137–139, 1982.

Koide, M., R. Michel, E.D. Goldberg, M.M. Herron, and C.C. Langway, Jr., Depositional history of artificial radionuclides in the Ross Ice Shelf, Antarctica, *Earth and Planetary Science Letters, 44,* 205–233, 1979.

Perkins, R.W., and C.W. Thomas, Worldwide Fallout, in *Transuranic Elements in the Environment,* edited by W.C. Hansen, pp. 53–82, 1980.

Reeh, N., H.B. Clausen, W. Dansgaard, N. Gundestrup, C.U. Hammer, and S.J. Johnson, Secular trends of accumulation rates at three Greenland stations, *Journal of Glaciology, 20,* 27–30, 1978.

PRESENT STATUS AND FUTURE OF LEAD STUDIES IN POLAR SNOW[1]

C. C. Patterson

Division of Geological and Planetary Science, California Institute of Technology, Pasadena, California

C. Boutron

Centre National de la Recherche Scientifique, Laboratoire de Glaciologie, Grenoble, France

R. Flegal

Marine Laboratory, California State University, Moss Landing, California

Abstract. Recent investigations of lead concentrations in polar snows carried out in the Caltech ultraclean laboratory using mass spectrometric isotope dilution techniques confirm earlier work showing lead to be ~1 pg/g in old (Holocene) Greenland ice.

These later studies also show that concentrations of lead in Antarctica are <4 pg/g in surface snows and ~1 pg/g in ancient ice. It has not yet been possible for investigators to reduce lead contamination associated with collection procedures to allowable low levels in Antarctic surface snow. It will be necessary to inaugurate new and better collection procedures to solve this problem.

Resume of Latest Findings

The concentration of lead in present day surface snow in Greenland lies in the range of 100 to 200 picograms per gram [Murozumi et al., 1969; Davidson et al., 1981]. Ng and Patterson in recent published work [1981] have confirmed earlier work of Murozumi et al. [1969] showing concentrations of lead to be about one pg/g in old Greenland ice (Holocene). In Greenland today about 99% of the lead in surface snow is industrial [Ng and Patterson, 1981]. In Greenland during prehistoric times (Holocene) about half the lead in snow came from silicate dusts while most of the rest probably came from volcanic emissions [Ng and Patterson, 1981] and seaspray. In Greenland, Davidson et al. [1981], have found that the Pb/Si ratio is 12 times higher in newfallen snow than in air above the snow, which is in accord with observations for Pb/Si ratios in rain, snow, and air at other remote locations [Ng and Patterson, 1981].

In the Antarctic the resolution of the situation awaits further reliable study. The concentration of lead in surface snow in Antarctica has been shown by Boutron and Patterson's latest work [1983] to be less than about 4 pg/g with the possibility that it could lie in the range of 2 to 3 pg/g, while work from Ng and Patterson [1981] and Boutron and Patterson [1983] shows lead in old ice to be about 1 pg/g, a little less than the upper limit reported by Murozumi et al. [1969], set by their lower limit of sensitivity, but considerably larger than their prediction of 0.03 pg/g. In that region the experimental constraints, known with certainty from the reliable studies cited, are: (a) the proportions of industrial lead cannot exceed the proportion of natural lead in surface snow today by more than a factor of 2 or 3 [Boutron and Patterson, 1982], (b) the proportion of volcanic and seaspray lead (global) amounts from Ng and Patterson [1981] must exceed the proportion of natural silicate dust lead in prehistoric snow [Murozumi et al., 1969; Cragin et al., 1976] by an order of magnitude, according to prehistoric global mass inventories and observed prehistoric total lead concentrations in old ice [Ng and Patterson, 1981; Boutron and Patterson, 1983], (c) total concentrations of lead in old ice are about one pg/g [Boutron and Patterson, 1983] and (d) the mean value of 300 (ng Pb kg^{-1}/ng Pb m^{-3}) for the washout factor for lead in rain and snow [Settle and Patterson, 1982] can be combined with the upper limit of lead concentration in surface Antarctic snow to estimate the upper limit for the concentration of lead in present day Antarctic air to be 10 pg/m^3. Since part of the lead in surface snow originates from dry deposition after the snow has fallen, this upper limit must be decreased by a fraction, which is unknown at present, but could amount to a 2-fold correction. This suggests a present-day concentration, integrated over the year, of about 5 pg Pb/m^3. It is possible that peaks of lead concentrations in the stratigraphic firn record associated with unusually large volcanic eruptions together with increases of lead concentrations in recent decades due to industrial inputs might be observed. Evidence for the existence of even more exotic natural sources of nonsilicate dust lead may be disclosed. Such observations remain to be made by reliable measurement.

Concordance With Latest Measurements of Lead Deposition Fluxes and Vertical Profiles of Lead in Oceans

The remarkable relationship observed at different geographic locations between precipitation and dry deposition input fluxes of lead to the oceans, the vertical lead concentration profiles at those locations, and prehistoric output fluxes of lead, are illustrated in Table 1. Fluxes are from Settle and Patterson [1982], SPE & NPE ocean profiles are from Flegal and Patterson's latest unpublished

[1] This article was presented in 1982. Some of the data and interpretations in it are now out dated.

TABLE 1. Observed Relation Between Soluble Pb Fluxes and Ocean Concentration Profiles

	South Pacific Easterlies	North Pacific Easterlies	North Pacific Westerlies	North Atlantic Westerlies
present eolian input fluxes (ng/cm^2 yr)	3	6	50	170
prehistoric eolian input fluxes (ng/cm^2 yr)	0.4	0.4	0.3	3
ocean concentrations (pg/g)				
0–100 m	2.5	11	14	34
100–800 m	2.5	8	11	32
800–2500 m	1.2	1.5	3.5	14
2500–5000 m	0.8	0.9	1.1	5
prehistoric sediment output fluxes (ng/cm^2 yrs)	4	4	3	30

work [Flegal and Patterson, 1982] while NPW & NAW ocean profiles are from Schaule and Patterson [1981a] and Schaule and Patterson [1981b]. This table is a summary from Flegal and Patterson [1982].

It has been shown that the residence time of lead in surface waters is a few years, in thermocline waters is about 20 years, and in deep waters is about 80 years [Schaule and Patterson, 1981a; 1981b]. The prehistoric net flux of soluble lead traveling from surface waters by passive sorption to primary producer algae, through primary consumer zooplankton, then down through the water column via fecal pellets [Burnett and Patterson, 1980], then out of the water to authigenic minerals in Pleistocene sediments was about 1/10th of the present day eolian input fluxes in the Northern Hemisphere Westerlies. It has been proved conclusively from the Greenland ice record that the combined annual lead precipitation and dry deposition flux in the Northern Hemisphere increased about 10-fold within the past 3000 years up to several centuries ago, and increased about 20-fold more within the last two centuries [Murozumi et al., 1969; Davidson et al., 1981; Ng and Patterson, 1981]. The short residence time of lead in seawater enables it to respond quickly to such a change in input fluxes, and this is precisely what the vertical record of lead in seawater shows did happen, with an outstanding correspondence linking variations of lead concentrations in existing surface and thermocline waters with variations in measured input and output fluxes of lead to and from those waters. The global marine record of lead input fluxes, concentrations of lead in water column, and output fluxes of lead thus experimentally verify the predictions of both the Greenland [Murozumi et al., 1969; Davidson et al., 1981; Ng and Patterson, 1981] and Antarctic [Murozumi et al., 1969; Ng and Patterson, 1981; Boutron and Patterson, 1983] records of lead in snow, confirm mass balance considerations of industrial emissions of lead to the atmosphere [Cragin et al., 1975; Boutron, 1980] which indicate that the major portion of these emissions are injected into the Northern Hemisphere Westerlies, and lend credence to the prediction, made on the basis of such mass balance considerations, that eolian lead fluxes in the Southern Hemisphere are much smaller than those in the North.

Today in Greenland the combined precipitation and dry deposition flux of lead obtained from Murozumi et al. 1969] and Davidson et al. [1981] is about 5 ng Pb/cm^2 yr. This is congruent with the marine global eolian lead flux picture, shown in Table 1, when it is considered that the circumpolar atmospheric circulation barrier hinders the migration of industrial lead northward from lower latitudes. In the Southern Hemisphere there is a precipitous decline in eolian lead fluxes in accordance with greatly diminished industrial lead emissions to the atmosphere in that hemisphere. Tentative lead in Antarctica snow data from Boutron and Patterson [1983] show an eolian flux of about 0.04 ng Pb/cm^2 yr for that region. The deposition and precipitation flux of lead in the South Pacific Westerlies has not yet been measured (to be done in the New Zealand SEAREX experiment). It may turn out to be less than that measured for the South Pacific Easterlies (<5 ng/cm^3 yr). If so, the estimated present-day deposition and precipitation flux of lead in the Antarctic may stand in about the same proportion to that in the South Pacific Westerlies (~0.04:2 or ~1:50) as the flux in Greenland is observed to stand to those in the North Pacific and North Atlantic Westerlies (5:150 or ~1:30).

Relation to Earlier Findings

Previous determinations of lead in Greenland snow and firn older than six decades made by all investigators except Murozumi et al., [1969] have been shown to be in gross error due to field and laboratory contamination [Ng and Patterson, 1981]. Analyses for lead in more recent snows (last 5 decades) in Greenland performed by all investigators except Murozumi et al., [1969] and Davidson et al., [1981] must now be considered compromised by field and laboratory contamination and therefore subject to an unknown amount of positive error [Boutron and Patterson, 1983]. This includes the analyses by Boutron [1979], Herron et al. [1977], Cragin et al. [1975], Gjessing [1977], and Jaworowski et al. [1981]. Studies of lead contamination on firn cores collected by workers wearing plastic gloves using plastic corers and stored in plastic tubes, all washed with HF and sealed in a clean laboratory for transport to the field, showed lead contamination of the outsides of those cores ranging from 30 to 3000 pg Pb/g [Boutron and Patterson, 1983]. This lead is readily transferred to the interior of the permeable cores if and when the outsides of the cores are subjected to any melting whatsoever. The degree of contamination at the interior of the cores by this exterior lead varies directly with the degree to which the exterior of the core was warmed and melted in the field. All of Boutron's firn cores collected under ultraclean field conditions from the Antarctic in 1981 are compromised to varying degrees by this effect and can yield only upper limit concentrations as a consequence. Less rigorously clean field sampling conditions in Greenland that have been carried out in the past [Boutron, 1979; Herron et al., 1977; Cragin et al., 1975; Gjessing, 1977; and Jaworowski et al., 1981] have evoked even greater contamination effects.

Lead concentrations are so high in Greenland snow laid down during recent decades that this contamination effect, assuming ultraclean procedures and precautions against melting were used and taken, would not amount to more than a 10% error. Such precautions were taken recently in Greenland only by Davidson et al. [1981], while Murozumi et al. [1969] used large 30 kg blocks of firn with low surface-to-mass ratios, and they did use ultraclean field procedures. Consequently for now, only those two studies of lead in Greenland snow laid down in recent decades can be accepted as scientifically significant. All others must be rejected because they are subject to an unknown amount of positive error. It nevertheless should be feasible to collect permeable firn cores by the ultraclean methods and then shave those cores in an ultraclean laboratory by methods carried out in Patterson's laboratory to obtain scientifically significant lead data when original internal core concentrations of lead are greater than 20–30 pg/g.

For lower concentrations of lead in older firn and ice in Greenland, the unavoidable core surface contamination added during drilling and transferred to the interior of the cores through cracks initiated by the drilling process compromises the samples beyond scientific usefulness. Such materials must be sampled through excavation by hand as large blocks with low surface-to-mass ratios under ultraclean field conditions. This can be done economically either by steam melting shafts [Patterson et al., 1970] or by excavating ice blocks from shallow pits dug below the perennial frost line in ablation zones.

In the Antarctic there are at present no scientifically acceptable measurements of lead concentrations in recent snows or air. The mean value of 15 pg Pb/g obtained by Murozumi et al. [1969] and used by Ng and Patterson [1981] to claim that lead in surface snow in Antarctica is 10-fold higher than in prehistoric times must now be rejected on the basis of the latest findings by Boutron and Patterson [1983]. The number used by Murozumi et al. [1969] is the mean of a mixture of several high outlier spikes (10 to 50 pg/g) with a larger number of less than detectable limit measurements (<1 to <4 pg/g). All other published data, such as Jaworowski et al., [1981], Boutron and Lorius [1979], Boutron [1980], Landy and Peel [1981], and Zhigalorskaya et al. [1974], and others must be rejected, since those reported values are even higher, and none of those investigators has demonstrated an ability to accurately and reliably measure and control in the field and laboratory lead contamination at the 0.1 pg/g level in water or at the 1 pg/m^3 level in air. A radically new approach must be taken in the field collection of permeable snow and firn in the Antarctic in order to obtain reliable lead concentration data. For old impermeable ice, samples must be collected as in Greenland, in the form of large blocks with low surface-to-mass ratios. They must be collected below the perennial frost line and not be fractured by shocks through sudden pressure release. Lead concentrations in Antarctic air reported in Murozumi et al. [1969] and Maenhaut et al. [1975] can no longer be accepted, since the best estimated concentration level is <10 pg/m^3, and a level of 30 pg/m^3 has now been reliably measured in the South Pacific Easterlies [Settle and Patterson, 1981].

Davidson et al. [1981] have shown that the Pb/Si ratio in air is enriched in snow and then subsequently reduced by dry deposition on the fallen snow. If this same process has operated during the past century in the Antarctic, then it may obscure volcanic pulses of lead recorded in firn stratigraphy, because the enrichment-depletion effect associated with precipitation and dry deposition is variable and subject to meteorological caprice. It therefore is important that accurate and reliable measurements in the Antarctic be simultaneously made of Pb/Si ratios in air, newfallen snow, and aged snow to evaluate the general characteristics of this effect.

Future Work

From the Roman era through the medieval period, variations in European base metal smelting productions with time are recorded in the Greenland ice record. The delineation of this archaelogical record should rank as an important goal in the NSF DPP. A suitable site should be chosen where this record is compressed as a consequence of low accumulation and yet is totally contained within impermeable ice, and samples excavated under ultraclean field conditions from tunnels and shafts.

The expected high proportion of volcanic and seaspray lead to silicate dust lead in Antarctic air older than several centuries provides a unique opportunity to study the historical record of volcanically induced pulses of heavy metals in the atmosphere, or, what may turn out to be even more scientifically important, other unknown exotic sources of nonsilicate dust lead. Here we are not constrained to Roman or medieval times and can go back to more logistically feasible periods. A series of ice blocks covering a 5000 year span can be excavated below the perennial frost line at an ablation zone dated by $^{18}O/^{16}O$ and ^{14}C. However, there is trouble. In Greenland, we would be dealing with about 10 pg Pb/g during Roman and medieval times, but in Antarctica concentration levels will be an order of magnitude less, and the spectre of capricious variations of precipitation Pb/Si enrichment and dry deposition Pb/Si depletion haunts the record. So measurements of Pb/Si in air, newfallen snow, and aged snow must be carefully made to disclose the magnitude of this possible obfuscating effect.

How should the qualifications of future investigators be evaluated? One criterion is that they must be able to measure 100 fg Pb/g water with a 10% error. They must be able to measure 1 pg Pb/m^3 air with a 10% error. Claims by workers using relative signal strength detectors that they can do this cannot be accepted at face value. They must be checked against IDMS methods. A second criterion is that they must be able to delineate in quantitative and accurate terms each of the many different kinds of lead contamination introduced to their samples from field apparatus, field operations, laboratory ware, reagents, and air to show that they can accurately measure and reliably control lead contamination at levels that are no more than about 10% of the expected lowest amounts of sample leads that will be encountered [Patterson and Settle, 1976; Patterson, 1982].

References

Boutron, C., Trace element content of Greenland snows along an east-west transect, *Geochimica et Cosmochimica Acta, 43,* 1253–1258, 1979.

Boutron, C., Respective influence of global pollution and volcanic eruptions on the past variations of the trace metals content of Antarctic snows since 1880's, *Journal of Geophysical Research, 85,* 7426–7432, 1980.

Boutron, C., and C. Lorius, Trace metals in Antarctic snows since 1914, *Nature, 277,* 551–554, 1979.

Boutron, C., and C. C. Patterson, The occurrence of lead in Antarctic recent snow, in firn deposited over the last two centuries, and in prehistoric ice. Describes analyses carried out 8/81 to 12/81 by Boutron and Patterson in Patterson's laboratory on samples collected by Boutron 1/81 to 2/81 from the Antarctic. *Geochimica et Cosmochimica Acta, 47,* 1355–1368, 1983.

Burnett, M. W., and C. C. Patterson, Perturbation of natural lead transport in nutrient calcium pathways of marine ecosystems by industrial lead, in *Isotope Marine Chemistry,* edited by E. Goldberg, Y. Horibe, and K. Saruhushi, pp. 413–438, Uchida Rokakuho Publishing Co., Tokyo, Japan, 1980.

Cragin, J. H., M. M. Herron, and C. C. Langway, Jr., The chemistry of 700 years of precipitation at Dye 3, Greenland, *Research Report 341,* U.S. Army Corps of Engineers, Cold Regions Research and Engineering Laboratory, Hanover, N.H., 1975.

Cragin, J. H., M. M. Herron, C. C. Langway, Jr., and G. A. Klouda, Interhemispheric comparison of changes in the composition of atmospheric precipitation during the late Cenozoic Era, in *Polar Oceans,* edited by M. J. Dunbar, pp. 617–631, Arctic Institute of North America, Calgary, Alberta, 1977.

Davidson, C. I., L. Chu, T. C. Grimm, M. A. Nasta, and M. P. Qamoos, Wet and dry deposition of trace elements onto the Greenland ice sheet, *Atmospheric Environment, 15,* 1429–1437, 1981.

Flegal, R., and C. C. Patterson, Vertical concentration profiles of lead in the Western Pacific at N 15° and S 20°, *Earth and Planetary Science Letters, 64,* 19–32, 1982.

Gjessing, Y. T., The filtering effect of snow, Symposium on Isotopes and Impurities in Snow and Ice (Proceedings of the Grenoble Symposium, 1975), *IAHS-AISH Publication No. 118,* 199–203, 1977.

Herron, M. M., C. C. Langway, Jr., H. V. Weiss, and J. H. Cragin, Atmospheric trace metals and sulfate in the Greenland ice sheet, *Geochimica et Cosmochimica Acta, 41,* 915–920, 1977.

Jaworowski, Z., M. Bysiek, and L. Kownacka, Flow of metals into the global atmosphere, *Geochimica et Cosmochimica Acta, 45,* 2185–2199, 1981.

Landy, M. P., and D. A. Peel, Short-term fluctuations in heavy metal concentrations in Antarctic snow, *Nature, 291,* 144–146, 1981.

Maenhaut, W., W. N. Zoller, R. A. Duce, and G. L. Hoffman, Concentrations and size distributions of particulate trace elements in the South Pole atmosphere, *Journal of Geophysical Research, 84,* 2421–2421, 1975.

Murozumi, M., T. J. Chow, and C. C. Patterson, Chemical concentrations of pollutant lead aerosols, terrestrial dusts and sea salts in Greenland and Antarctic snow strata, *Geochimica et Cosmochimica Acta, 33,* 1247–1294, 1969.

Ng, A., and C. C. Patterson, Natural concentrations of lead in ancient Arctic and Antarctic ice, *Geochimica et Cosmochimica Acta, 45,* 2109–2121, 1981.

Patterson, C. C., T. J. Chow, and M. Murozumi, The possibility of measuring variations in the intensity of world wide lead smelting during medieval and ancient times using lead aerosol deposits in polar snow strata, in *Scientific Methods in Medieval Archaeology,* edited by R. Berger, pp. 339–350, University California Press, 1970.

Patterson, C. C., and D. M. Settle, The reduction of order of magnitude errors in lead analyses of biological materials and natural waters by evaluating and controlling the extent and sources of industrial lead contamination introduced during sample collecting and analysis, Accuracy in Trace Analysis, in *National Bureau of Standards Special Publication, No. 422,* edited by P. La Fleur, pp. 321–351, 1976.

Patterson, C. C., Analyses for lead at levels of 10^{-13} to 10^{-9} g/g, invited manuscript, 1982, rejected by editors of Journal of Association of Official Analytical Chemists, submitted by invitation to *Analytical Chemistry.*

Schaule, B. K., and C. C. Patterson, Lead concentrations in the Northeast Pacific: evidence for global anthropogenic perturbations, *Earth and Planetary Science Letters, 54,* 97–116, 1981a.

Schaule, B. K., and C. C. Patterson, Perturbations of the natural lead depth profile in the Sargasso Sea by industrial lead, Trace Metals in Sea Water, *Proceedings of the NATO Advanced Research Institute,* 1981b, Plenum Press, 1983.

Settle, D. M., and C. C. Patterson, Unpublished data from the American Samoa SEAREX experiment in the South Pacific Easterlies, 1981.

Settle, D. M., and C. C. Patterson, Magnitudes and sources of precipitation and dry deposition fluxes of industrial and natural leads to the North Pacific at Enewetak, *Journal of Geophysical Research, 87,* 8857–8869, 1982.

Zhigalorskaya, J. N., Z. P. Makhon'ko, A. I. Shishkina, V. V. Egorov, C. C. Malakhov, and R. I. Pervunina, Content of some miro elements in natural waters, *Trdy Institut Eksperimental-noi Meteorologii, 2,* 86–113, 1974.

A GEOPHYSICAL SURVEY OF SUBGLACIAL GEOLOGY AROUND THE DEEP-DRILLING SITE AT DYE 3, GREENLAND

K. C. Jezek[1], E. A. Roeloffs, and L. L. Greischar

Geophysical and Polar Research Center, University of Wisconsin-Madison, Madison, Wisconsin

Abstract. A geophysical survey of subglacial geology was carried out upstream of the Dye 3 borehole in order to characterize bedrock morphology and near-surface crustal structure in an area believed to be close to the boundary between the Archaean craton and the Nagssugtoqidian mobile belt. The survey consisted of radar determinations of bedrock elevation and gravity measurements to be used in estimating near-surface densities.

Free-air anomalies are strongly positive (about + 80 mgal) and are about 50 mgal higher than the 1° × 1° mean free-air anomalies determined by satellite measurements. Results of our gravity surveys west of Dye 3, as well as along the EGIG line, suggest the positive anomalies are indicative of an eastward increasing trend in gravity across southern Greenland, perhaps due to regional structure in the crust and mantle rather than to glacial isostasy.

The complex bedrock topography in the Dye 3 area was mapped using radar operated from the surface. The effect of bedrock topography on the free-air gravity was modeled in three dimensions using a program based on the method of Talwani and Ewing. The uniform density model of bottom structure that gives the best fit to the gravity data has a density of 2.7 g/cc and is morphologically similar to the bedforms found in the more extensive survey presented by Overgaard and Gundestrup. The bedrock topography map we show represents our best estimate after combining the radar data with our 2.7 g/cc gravity model. We find a rugged, at times precipitous, terrain whose features have a wavelength of about 8 km, an amplitude of 350 m and a roughly north-south strike. Although the terrain suggests that this area has been structurally altered, perhaps by tectonic forces, the data provide no direct means of distinguishing between the Archaean block and the mobile belt. Morphological differences between our middle site and Dye 3 weakly suggest a geologic boundary may lie between the sites.

Introduction

A geophysical survey, consisting of detailed radar soundings of bedrock morphology, gravity mapping, and surface elevation measurements, was made around the deep drilling site at Dye 3 during the summer of 1981. The measurements were made in conjunction with the Ohio State University glaciological program [Whillans et al., 1982] to map glacial flow upstream of the borehole using surface strain-rate data. While part of our effort was directed toward supplementing radar data collected by the Technical University of Denmark party (see Overgaard and Gundestrup [1984]) along Whillans' strain network, the bulk of our studies have been aimed at answering questions about subglacial geology in an area postulated to lie near the boundary between the Greenland portion of the North Atlantic Archaean craton and the Nagssugtoqidian mobile belt (Figure 1). In addition to Dye 3, surveys were carried out at the two strain figures to the west, shown in Figure 1. Only preliminary results from these western survey sites are presently available.

Glaciological and Geological Setting

The deep drilling site at Dye 3 is located at lat 65.2°N, long 43.8°W, approximately 40 km east of the ice divide that strikes roughly NW-SE in this part of Greenland. The average ice thickness at Dye 3 is about 2 km, and ice flow is to the northeast. The Greenland ice sheet is considered to be relatively stable, since much of it rests on bedrock that would be above sea level were the ice sheet removed, and reconstructions of the ice sheet during the Wisconsin maximum [Hughes et al., 1981] suggest that at most several hundred meters of thinning have occurred in the Dye 3 area since the onset of glacial retreat.

Very little is known about the bedrock geology beneath the Greenland Ice Sheet. Based on the correlation of eastern and western coastal outcrops, the Dye 3 area is postulated to lie near the northern boundary of the Greenland portion of the North Atlantic Archaean craton, which is bordered to the north by the Nagssugtoqidian mobile belt. A summary of the geology of the Archaean coastal outcrops can be found in Bridgwater, Watson, and Windley [1973]. The rocks of which the Archaean craton is composed are mainly gneisses of amphibolite and granulite facies, interleaved with supracrustal rocks including amphibolites, mica schists, marbles, and quartzites.

The Archaean structural and metamorphic history of the craton contains two episodes that might contribute to present day surface morphology and density variations. The first event was an "interleaving" event, of disputed mechanism, which served to produce concordant contacts between basement gneisses, supracrustal rocks, and intrusive bodies of widely different ages. The second event was a regional high grade metamorphism at about 2900 Ma that produced a tectonic style characterized by dome and basin structures and an absence of linear grain. During the period from 2700 to 1700 Ma, the Archaean craton stabilized, but activity began in mobile belts bordering the craton. Diabase dike swarms were intruded into the craton and the bordering mobile zones. In particular, the dikes emplaced in the Nagssugtoqidian belt were deformed into parallelism with country rock structures. This change in degree of deformation of the diabase dikes defines the boundary between the Archaean craton and the Nagssugtoqidian belt. The rocks of the

[1] Now at: U.S. Army, Cold Regions Research and Engineering Laboratory, Hanover, New Hampshire.

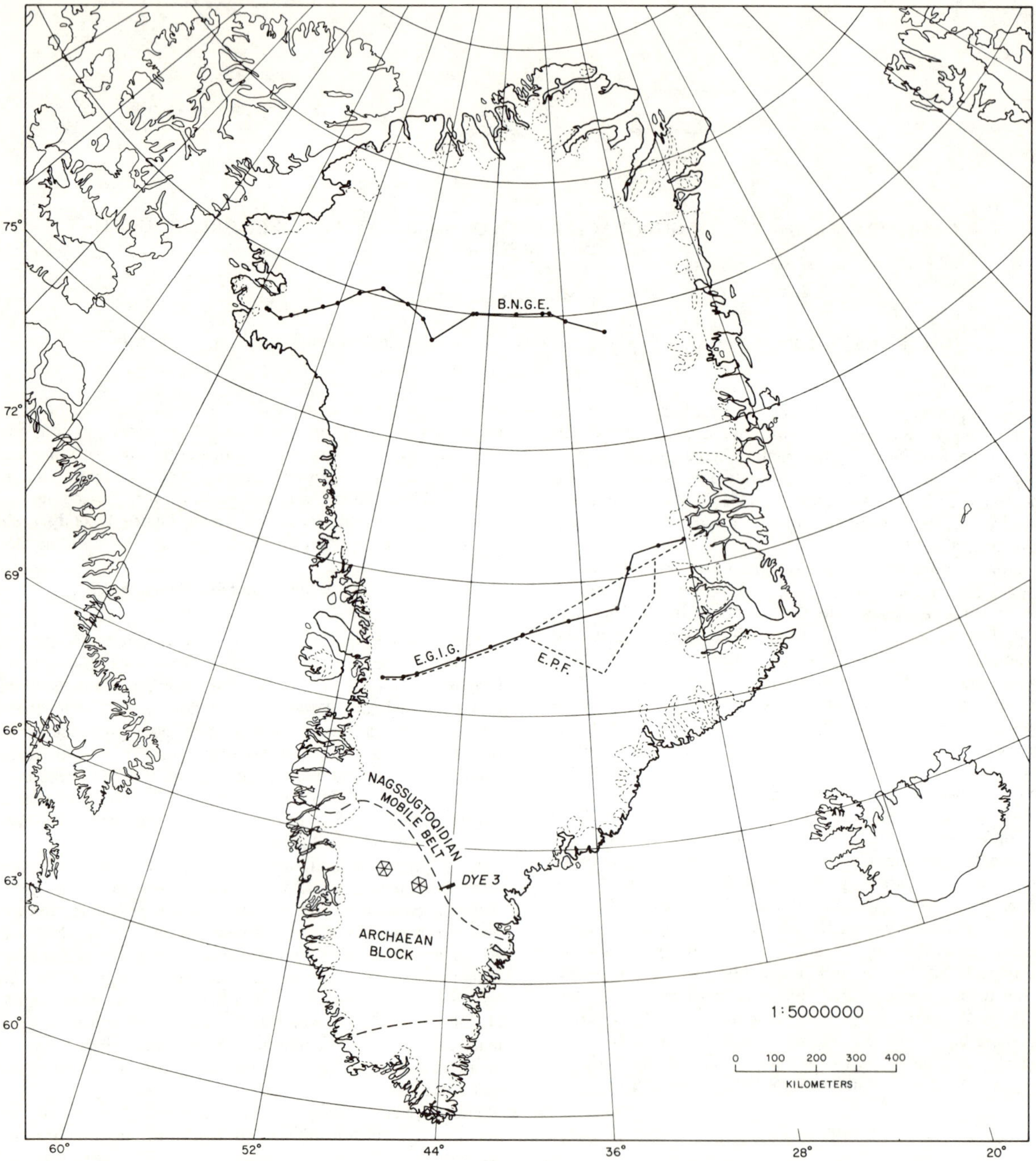

Fig. 1. Map of Greenland showing the Whillans' polygonal strain figures along which radar and gravity data were collected. Shaded area labeled DYE 3 shows the extent of the local survey near Dye 3. Earlier traverses in Greenland along which gravity observations were made are also shown.

Nagssugtoqidian belt are mainly reworked Archaean rocks with a tectonic style that has been called a "large-scale augen structure", in which pod-shaped zones of relatively undisturbed Archaean rocks occur within more intensely deformed zones possessing a linear structural grain that parallels the boundary of the craton [Bridgwater, Escher, and Watterson, 1973].

Field Procedure

The data described in this paper were collected along the lines shown in Figure 3. Radar data were collected continuously along each of these lines; gravity measurements were made at the flagged points shown in Figure 4. Absolute distances are accurately known

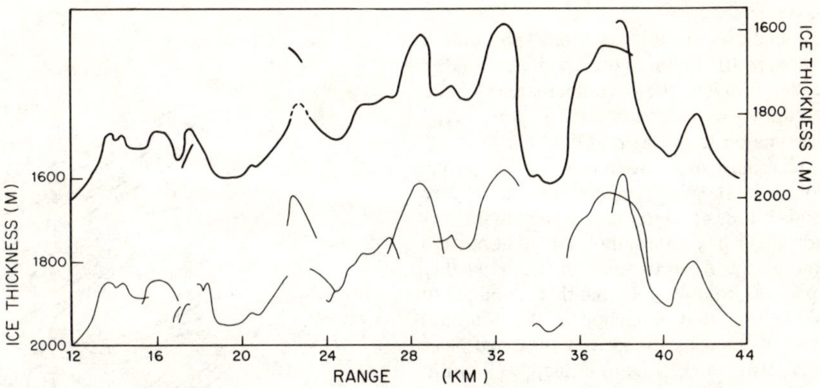

Fig. 2. Unmigrated (lower curve) and migrated (upper curve) ice thickness sections along the A line. Range is distance from Dye 3.

along the first 20 km of the A, B, and C lines since they were determined by laser-ranging devices (Whillans, personal communication), while distance measurements along the extension of the A line and along the D line were less precisely determined by vehicle odometer.

Radar measurements were made from the surface using a SPRI Mark II radar unit operating at 35 MHz, having a pulse width of 200 ns and a system response of 153 dB. Individual folded dipole antennas, one-half wavelength long and having a reflecting element located one-quarter wavelength above the dipole, were used for the transmitter and receiver and were spaced 31 m apart. An intensity-modulated display of the return waveform was recorded on a Honeywell oscillograph. In order to correlate the oscillograph record with position on the snow surface, a mark was placed on the record and the time was recorded as each flag was passed. Vehicle speeds computed from these data were nearly constant, and imply a position uncertainty of ± 30 m. Along with continuous profiling, Polaroid photographs were taken of the received waveform displayed on the oscilloscope screen. A 1 MHz timing calibration signal was routinely placed both on the Polaroid pictures and the oscillograph output.

A LaCoste and Romberg Model G geodetic gravimeter (G19) was used for the gravity survey. Each field observation consisted of at least two consecutive readings generally reproducible to a few hundredths of a milligal. Each series of observations formed a closed loop tied to a base station established on the base of the southeastern column supporting the Dye 3 station. (A second base station was situated away from Dye 3 at Whillans' satellite station 3003, lat 65.2°N, long 43.8°W.) The survey was tied to the International Gravity Standardization Net 1971 (IGSN-71) from several IGSN 71 stations in Sondrestromfjord (primarily IGSN 71 984365).

Satellite-derived absolute elevation measurements relative to the geoid were provided at several points on the survey (Whillans, personal comm.) and elevation measurements between these sites were carried out using hand-held altimeters (Gundestrup, Herron and Whillans, personal communication). Comparison between overlapping points on lines surveyed by the different groups suggest a relative elevation accuracy of ± 2 m corresponding to an accuracy of ± 0.6 mgal in the free-air gravity anomaly. We believe the absolute elevation accuracy to be approximately ± 2 m as well.

Radar Data Reduction

Raw travel times from the oscillograph records were converted to equivalent raw ice thicknesses (not necessarily vertical thickness) using an average wave speed of 171 m/μs. The raw ice-thickness data show numerous diffraction hyperbolas and possible side echoes, suggesting a complex bottom topography. In order to exploit the data to the fullest extent, we have developed and applied a simple two-dimensional migration algorithm, similar to the method described by Harrison [1970], for removing diffracted arrivals and correcting bottom slopes to their true inclinations. A good description of the need for applying two-dimensional migration in a complex area (such as Dye 3) is given by French [1974] who also compares the results of two and three-dimensional migration. We point out two reasons why the two-dimensional approach is valid in this case. First, migrating in two dimensions will properly orient side scatterers along the profile line although there will still be cross track uncertainty. Second, the pair of antennas has a beam that is oriented more along the track than across the track. We have used data collected on a zone of bottom crevasses in the Ross Ice Shelf [Jezek et al., 1979] to show that the ratio of the range seen by our antenna array along the traverse path to the range seen across the traverse path is approximately 3/2. At Dye 3 we detected targets along the track at ranges of at least 800 m. Therefore a maximum cross-track uncertainty is ± 500 m.

Raw and migrated profiles along line A are shown in Figure 2. The data are improved by the migration routine although some multiple echoes remain (for instance at range mark 38) indicating side echoes are present.

Data Reduction—Gravity

Gravity anomalies were computed using the datum of the International Gravity Standardization Net (IGSN 71) and relative to the gravity formula of the Geodetic Reference System, 1967. Free-air corrections were applied to all the data and the free-air anomaly is contoured in Figure 4. Additional corrections to the gravity data were complicated by surface and bedrock topographic effects, and while surface variations were smoothed to the 2500 m contour level using a simple Bouguer correction, the bottom topographic variations were too irregular for a similar correction to be made. So, three-dimensional modeling of the effect of bottom topography on the free-air gravity data was done using the approach described by Talwani and Ewing [1960] that approximates bottom topography by a group of polygons defined by contours, in this case, spaced 100 m apart. Adjustments were made to the model density and more importantly to the variations in topography around the perimeter of the radar survey. Because of the uncertainty in the behavior of bottom topography outside of the range of our radar survey, we assumed that all gravity anomalies were due to topography unless the change in topography needed to eliminate the anomaly violated the radar data. As the model evolved, the perimeter areas of the survey were changed and redrawn so that the gravity field predicted by the model closely approximated the measured field.

We have refined the radar map of bottom topography into a topographic model that closely fits the observed gravity data (Figure 3). The map reveals a rugged, at times precipitous, terrain characterized by rounded knolls and valleys having an amplitude of about 350 m and wavelength of about 8 km. A north-south strike is weakly suggested by some of the patterns, and we note that this is oblique to the ice flow direction. We believe that extension of the contours beyond the survey lines was justified by the gravity data and have learned, subsequent to our own work, that our map of bottom topography is generally similar to the more extensive radar survey by Overgaard and Gundestrup [1984]. Among uniform near-surface crustal densities, a value of 2.7 g/cc gave the best fit to the gravity data—2.6 or 2.8 g/cc yielded significantly worse fits.

It is necessary to add the regional free-air anomaly to the model free-air gravity before the model values and measured values can be compared. Since the magnitude of the regional anomaly is unknown, a uniform offset of +58 mgal, determined by applying a least-square fit to the difference between the measured and modeled gravity, was added to the modeled gravity values. The magnitude of this offset is equal to the topographically corrected free-air anomaly.

Model values and measured free-air anomalies are compared directly in Figure 5 and the residual gravity difference between model and observed values are contoured in Figure 6. It is likely that most of the residuals greater than 1 mgal can be attributed to topographic uncertainties. For example, the large −2 mgal anomaly at the SW end of the D line is likely to be associated with the steep-sided valley identified by radar on the A line. Assuming that the valley extends north toward the D line accounts for some but not all of this gravity anomaly. The only anomaly that may reflect a real density variation is the high near stations B5-B7 and C5, since the good radar coverage in the area makes it unlikely that the topographic high in this area is broader and flatter than was modeled.

The residual anomaly map (Figure 6) can be considered as the variation about the mean of the Bouguer anomaly. Topographic variations above the 250 m elevation contour have been replaced by ice assuming a density contrast of 1.8 gm/cc. The mean Bouguer anomaly found after removing 2250 m of ice and 250 m of rock is −54 mgal.

The new gravity data presented here may represent the best reduction of gravity observations in Greenland because it was possible to remove three-dimensional topographic effects. Previous gravity surveys on the Greenland Ice Sheet have included minimal lateral coverage.

Discussion

The Dye 3 survey has revealed several interesting facts about the Greenland crust at this location: (1) the free-air anomalies are strongly positive (+ 80 mgal); (2) near-surface crustal density is 2.7 gm/cc; (3) Bouguer anomalies are on average −54 mgal; (4) the bottom topography is very rugged. We will discuss these results in the context of previous surveys in Greenland (the International Greenland Expedition (EGIG) line [Brockamp, 1965], the British North Greenland Expedition (BNGE) line [Bull, 1955], and the Expedition Polaires Francaises (EPF) line [Martin et al., 1954]), and also in relation to preliminary results from our western survey sites.

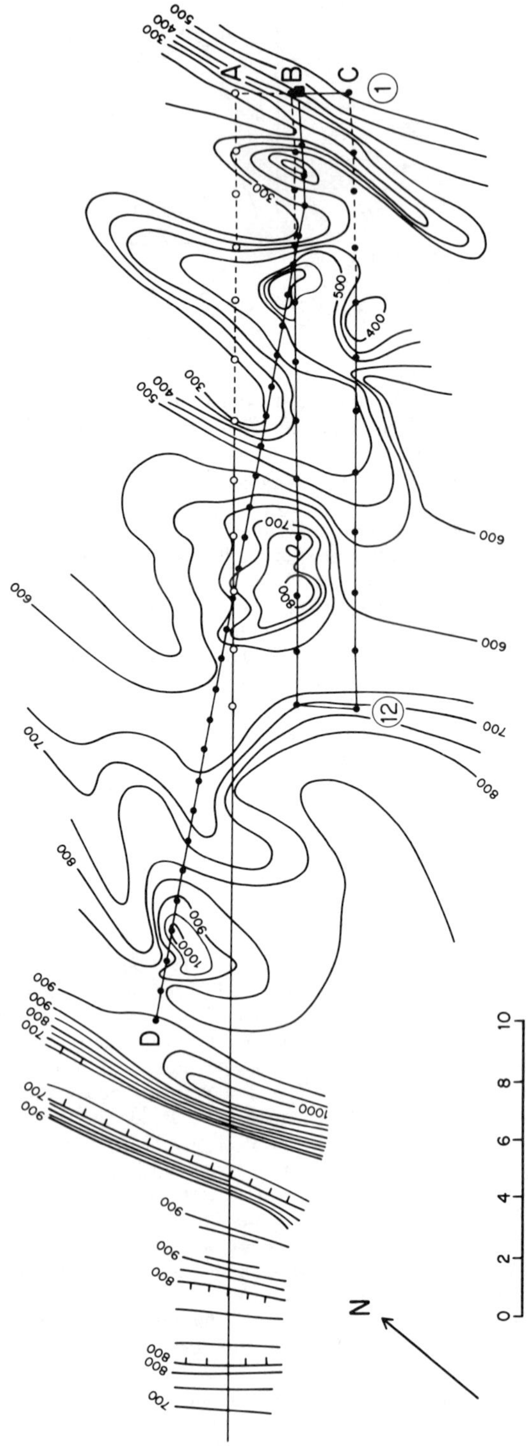

Fig. 3. Bedrock topography compiled from radar and gravity data. The straight solid lines which connect dots show radar profiles along which good bottom reflections were obtained. Dashed lines represent areas of poor bottom reflection. Contour interval is 50 m. The location of Dye 3 is shown by the dark square.

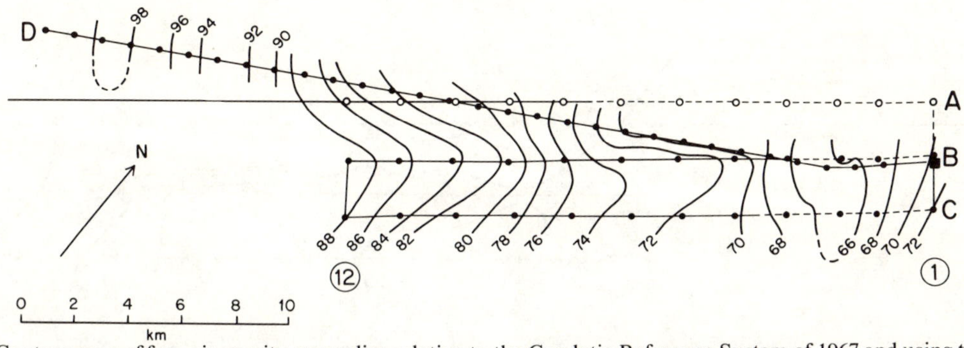

Fig. 4. Contour map of free-air gravity anomalies relative to the Geodetic Reference System of 1967 and using the datum of the International Gravity Standardization Net of 1971. Contour interval is 2 mgal. Solid circles indicate flagged positions at which gravity readings were taken. Open circles indicate positions used only for navigation.

The average free-air anomaly of +80 mgal is unusual. Preliminary analysis of the gravity data collected at base stations at our other survey sites suggests there is an eastwardly increasing trend (the free-air anomaly at 45.6° west longitude, 65.0° north latitude is +76 mgal and the anomaly at 47.5° west longitude, 65.3° north latitude is +47 mgal). There is a similar trend in the free-air anomalies along the EGIG line, but no eastward increase of free-air anomalies was observed by Bull [1955] in northern Greenland.

It is generally assumed that the ice sheets have thinned since the end of Wisconsin time—a process that would reduce free-air anomalies. Pawlowicz [1969] has suggested a rapid shrinkage of the southern Greenland ice sheet about 8,000 years ago, accompanied by isostatic uplift, and followed by rapid postsurge build up of the ice sheet for which isostatic compensation is not yet complete. However, it is difficult to understand how reaccumulation of the ice sheet could have been rapid enough to prevent isostatic readjustment. Furthermore, glacial reconstructions by Hughes, et al., [1981], do not suggest that there have been any dramatic changes of the inland Greenland ice sheet since the Wisconsin maximum. Therefore we prefer to hypothesize a geologic cause for the positive free-air anomalies. The trend of eastwardly increasing gravity values on our survey and on the EGIG line is consistent with the GEM 10 satellite gravity model [Lerch et al., 1977] and suggests that the large positive free-air anomaly may have a deep-seated origin. Because we suspect that the gravity anomalies may be due in part to isostatic imbalance in the earth, we do not make an estimate of crustal thickness based on the Bouguer anomaly as was done by Pawlowitz [1969].

The density of 2.7 gm/cc which produced the best gravity model falls within the range of densities inferred from seismic velocities [Brockamp, 1965] measured in central Greenland. There is no evidence of significant low-density morainal deposits in the valleys of the survey area.

Finally, we note again the rugged bottom topography revealed by our survey. Qualitative analysis of the data at our other survey sites indicates that bottom topography is much smoother at the central site and at the westernmost site. The variation of bottom topography between the survey sites can be only a weak clue indicating whether or not the rocks are associated with the Archaean block or mobile belt. While the bedrock terrain may be suggestive of the basin and dome features found in undeformed Archaean outcrops, it is not possible to distinguish between the main Archaean craton and the large, relatively undeformed, "augen" structures found in the mobile belt. Therefore we simply postulate that a boundary between two geologic provinces may exist between our central survey site and Dye 3.

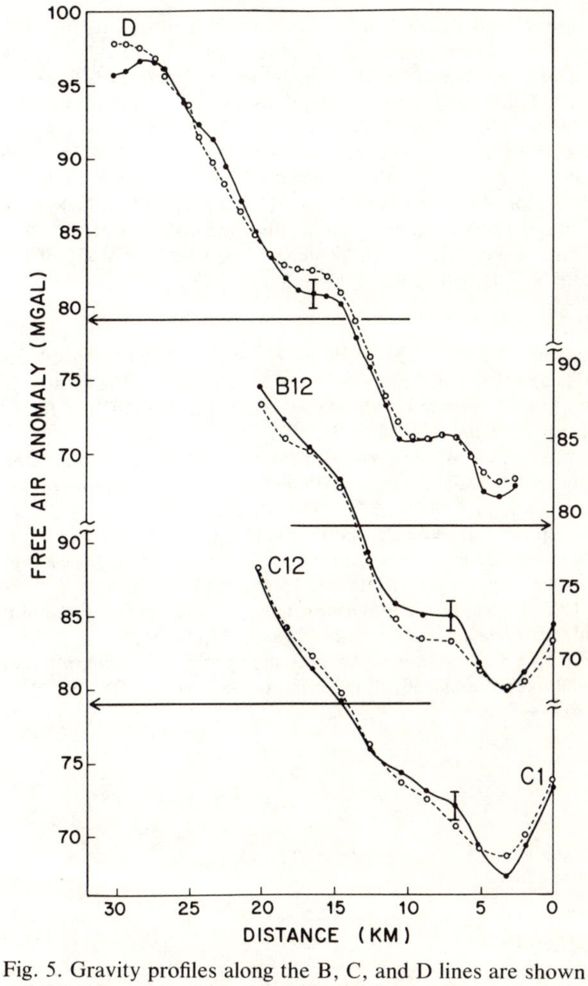

Fig. 5. Gravity profiles along the B, C, and D lines are shown with solid circles representing the observed free-air anomalies and open circles representing the modeled gravity anomalies. Large arrows are drawn at the value of the mean free-air anomaly. Error bars reflect both the accuracy of the free-air gravity observations and an estimate of the uncertainty involved in the topographic model. They are drawn at points where the difference between the observed and modeled gravity may be significant.

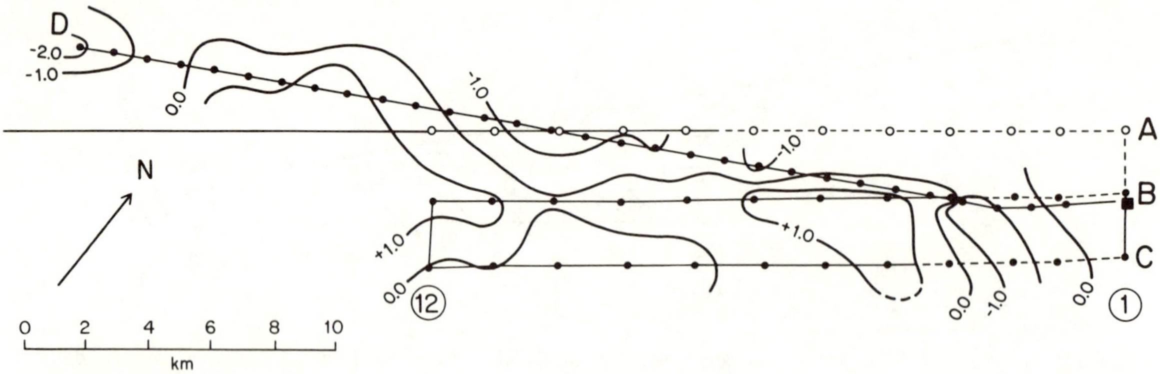

Fig. 6. Gravity anomaly residuals between the observed and modeled free-air gravity anomalies. Contour interval is 1 mgal.

Acknowledgments. We wish to thank I. Whillans for inviting us to join his field program and all the members of the 1981 field party for their assistance. We thank N. Gundestrup, S. Overgaard, C. Bentley and C. Hammer for many helpful discussions. Data used in this paper were collected while Jezek was a postdoctoral fellow at the Institute of Polar Studies of the Ohio State University. This research was supported by the National Science Foundation grants DPP78-20953 (Geophysical and Polar Research Center) and DPP80-08356 (Institute of Polar Studies). This is contribution #404 of the Geophysical and Polar Research Center and contribution #451 of the Institute of Polar Studies.

References

Bridgwater, D., A. Escher, and J. Watterson, Tectonic displacements and thermal acticity in two contrasting Proterozoic mobile belts from Greenland, *Philosophical Transactions of the Royal Society, London, 273,* 513–533, 1973.

Bridgwater, D., J. Watson, and B. F. Windley, The Archaean craton of the North Atlantic region, *Philosophical Transactions of the Royal Society, London, 273,* 493–512, 1973.

Brockamp, B., Uber einige geophysikalische Ergebnisse der internationalen Gronland-Expedition EGIG 1959, *Polarforschung, Band VI,* Jahrgang 35, Heft 1/2, 42–66, 1965.

Bull, C.B.B., Values of gravity on the inland ice of North Greenland, *Meddelelser om Gronland, Band 173,* 11, 1955.

French, W.S., Two-dimensional and three-dimensional migration of model-experiment reflection profiles, *Geophysics, 39,* 265–277, 1974.

Harrison, C.H., Reconstruction of subglacial relief from radio-echo sounding records, *Geophysics, 35,* 1099–1115, 1970.

Hughes, T.J., G.H. Denton, B.G. Anderson, D.H. Schilling, J.L. Fastook, and C.S. Lingle, The last great ice sheets: a global view, in *The Last Great Ice Sheets,* edited by D.J. Denton and T.J. Hughes, pp. 263–317, J. Wiley and Sons, New York, 1981.

Jezek, K.C., C.R. Bentley, and J.W. Clough, Electromagnetic sounding of bottom crevasses on the Ross Ice Shelf, Antarctica, *Journal of Glaciology,* 24 (90), 321–330, 1979.

Lerch, F.J., S.M. Klosko, R.E. Laubscher, and C.A. Wagner, Gravity model improvement using GEOS-3 (GEM 9 & 10), Preprint X-921-77-246, Goddard Space Flight Center, Greenbelt, Maryland, 1977.

Martin, J., P. Stahl, F. Munk, and A. Joset, Groenland 1948–1952, Gravimetrie 1ere partie, *Expeditions Polaires Francaises,* Paris, 152, 1954.

Overgaard, S., and N.S. Gundestrup, Bedrock topography of the Greenland Ice Sheet in the Dye 3 area, this volume, 1984.

Pawlowicz, E.F., An isostatic study of northern and central Greenland based on gravity values and airborne radar ice-thickness measurements, Ph.D. thesis, 98 p., the Ohio State University, 1969.

Talwani, M., and M. Ewing, Rapid computation of gravitational attraction of three-dimensional bodies of arbitrary shape, *Geophysics, 25,* 203–225, 1960.

Whillans, I.M., A.J. Drew, K.C. Jezek, and N. Gundestrup, Ice flow leading to the GISP borehole (abstract), *EOS Transactions, American Geophysical Union, 63,* 1982.

A STEADY-STATE PREDICTION OF DYE 3 CORE FEATURES

B. McInnes[1] and U. Radok

Cooperative Institute for Research in Environmental Science, National Oceanic and Atmospheric Administration,
University of Colorado, Boulder, Colorado

Abstract. The assumptions of a steady-state ice sheet with the present shape of Greenland and exact balance of its present net accumulation by ablation and calving have been used to simulate the flow and temperature fields upstream of and including the DYE 3 borehole site. Results include the vertical temperature and age profiles in the borehole and the points at which the ice at different depths in the core would have been deposited if the steady-state hypothesis were true.

Introduction

A deep ice coring project nowadays can be prepared by detailed radar surveys of the surroundings of the proposed site which tends to be chosen for access to old ice as well as simple base rock topography and current ice flow conditions. A possible further preparatory step that has proved useful more generally in the planning of antarctic programs is to derive the physical characteristics of the field-work region from the simplest model assumption—that the ice sheet is in a state of exactly zero net mass balance. The ice-flow velocities and temperatures then directly derivable from the equations of continuity and heat conduction provide standards of comparison for the observations and can also be used to test the model sensitivity to inadequate or erroneous input data.

The "balance" approach was pioneered by W. Budd for Antarctica [Budd et al., 1970] and has recently been applied to Greenland by Budd et al. [1982] and Radok et al. [1982]. In the last of the studies B. McInnes developed computer procedures that isolate and enlarge any chosen area of an ice sheet for closer analysis and display of its physical features. This procedure has been applied to a region measuring 400 km × 300 km around Dye 3, where in 1981 the second penetration of the Greenland ice sheet was achieved by the Danish, American, and Swiss glaciologists of the Greenland Ice Sheet Program (GISP). The "balance" estimates of ice velocities, temperatures, and ages are presented here as a contribution to the interpretation of the new measurements. It should be emphasized that as details extracted from a study of the ice sheet as a whole they provide smoothed orders of magnitude rather than precise local values.

Procedures

The procedures used have been described in detail in the Greenland study [Radok et al., 1982]. Briefly, the information available at the time on the ice sheet surface and bedrock elevations, surface mass balance, and surface layer (10 m) temperature was converted to smoothed fields represented by the 150 × 100 points of a grid with 20 km spacing covering Greenland. The upper part of Figure 1 shows a perspective view of the resulting surface shape as seen from the southeast. The grid portion around Dye 3 singled out for closer study is shown in the lower part of Figure 1. The grid flowline is made up of the grid points closest to a line constructed perpendicular to the surface elevation contours. For the age and temperature calculations the flowline input data along the portion upstream of Dye 3 were interpolated at 2 km intervals.

The contoured input data are shown in Figure 2 where the dots mark the locations of the measurements used. It is immediately apparent that far more information was available for the "topographic" variables, surface elevation and ice thickness, than for the "regime parameters", mass balanced and surface layer temperature. The implications will be taken up later.

The continuity equation for the steady state of zero net mass balance in flowline coordinates (m longitudinal, n transverse) has the form

$$\frac{\partial \bar{V}D}{\partial m} + \bar{V}D\frac{\partial \sin\psi}{\partial n} - A = 0 \qquad (1)$$

where

$\bar{V}$ is the vertically averaged ice flow velocity,
D is the ice thickness,
A is the mass balance, and
ψ is the angle between the flowline and one of the grid coordinates.

The second term arises from the divergence or convergence of the flowlines. The integration of equation (1) along each flowline yielded the balance isotachs shown in Figure 3; the flow direction everywhere being perpendicular to the surface elevation contours (Figure 2, upper left).

In the absence of topographic, regime, and climatic changes, the ice-flow velocity and accumulation rate determine the time change in the surface-layer temperature, T_s, of an ice column flowing from the center to the edge of the ice sheet. The lowering of the surface elevation, E, experienced by such a column creates in the surface layer a vertical temperature gradient [Robin 1955]

$$\gamma_s = -\frac{V_0\alpha\lambda}{A} \qquad (2)$$

[1] Now at: Meteorology Department, University of Melbourne, Australia.

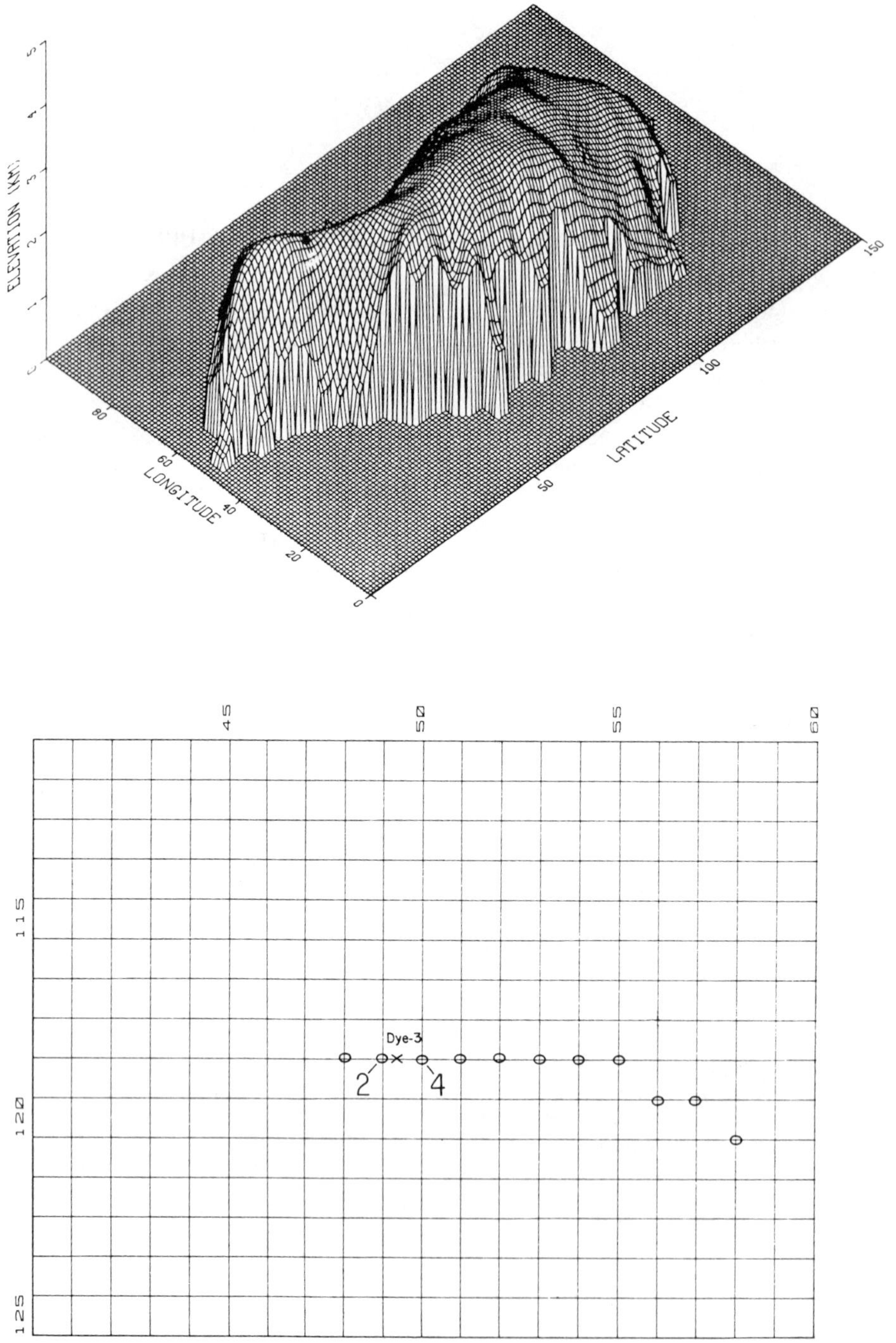

Fig. 1. Grid representations of the Greenland ice-sheet surface (top) and of the region around Dye 3 (bottom).

where α is the surface slope and the significance of $\lambda = \partial T_s/\partial E$; the "topographical lapse rate," has been discussed by Jenssen and Radok [1982]. Equation (2) constitutes the upper boundary condition for the solution of the thermal equation,

$$\frac{\partial T}{\partial t} + \underline{\mathbf{V}} \cdot \nabla T - q = K \frac{\partial^2 T}{\partial z^2} - w \frac{\partial T}{\partial z} \tag{3}$$

where

$T\ (x,y,z,t)$ is the ice temperature,
q is heat input at the base,
w is the vertical ice velocity,
$\underline{\mathbf{V}}$ is the horizontal ice velocity at the level considered, and
K is the thermal conductivity of the ice.

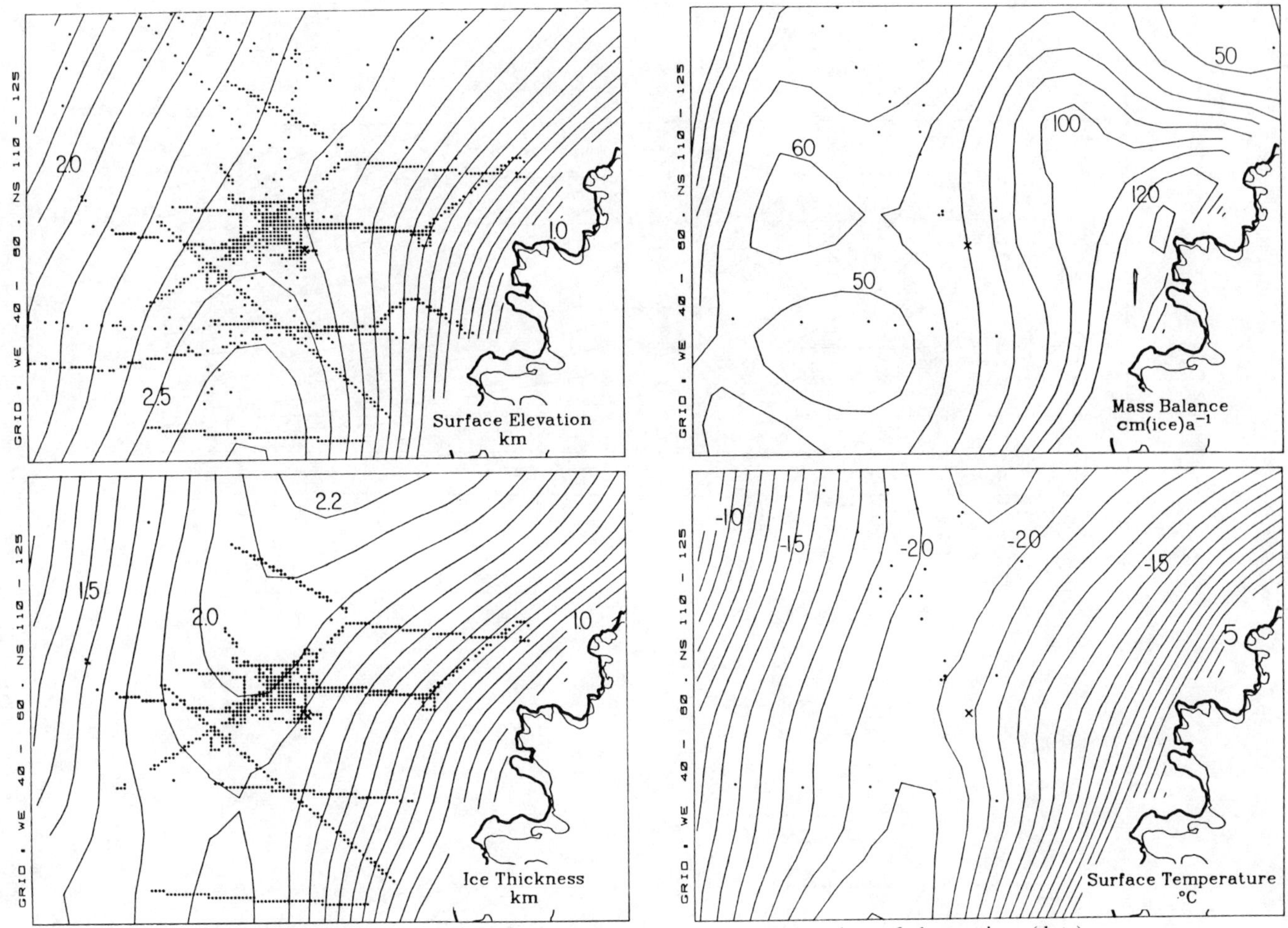

Fig. 2. Smoothed input fields for the steady-state model, and locations of observations (dots).

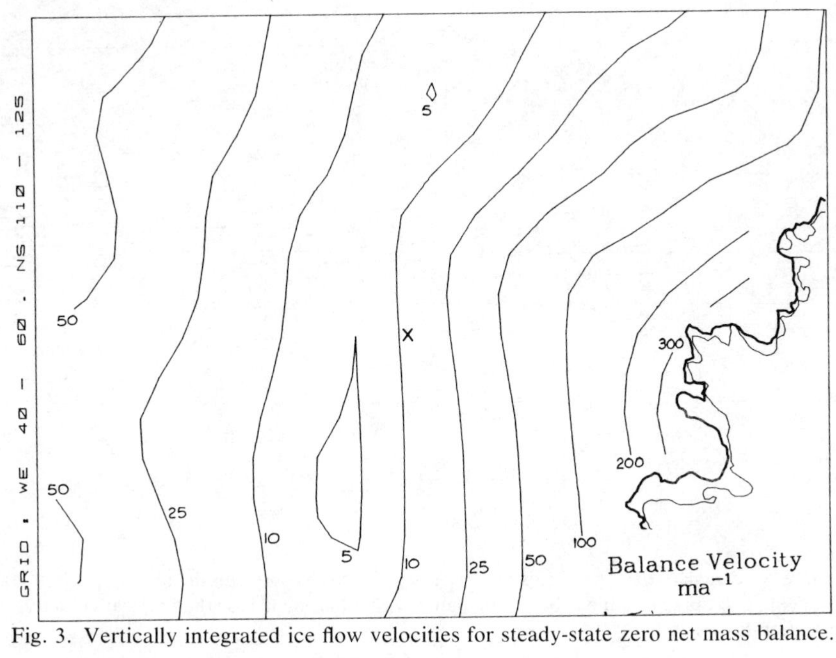

Fig. 3. Vertically integrated ice flow velocities for steady-state zero net mass balance.

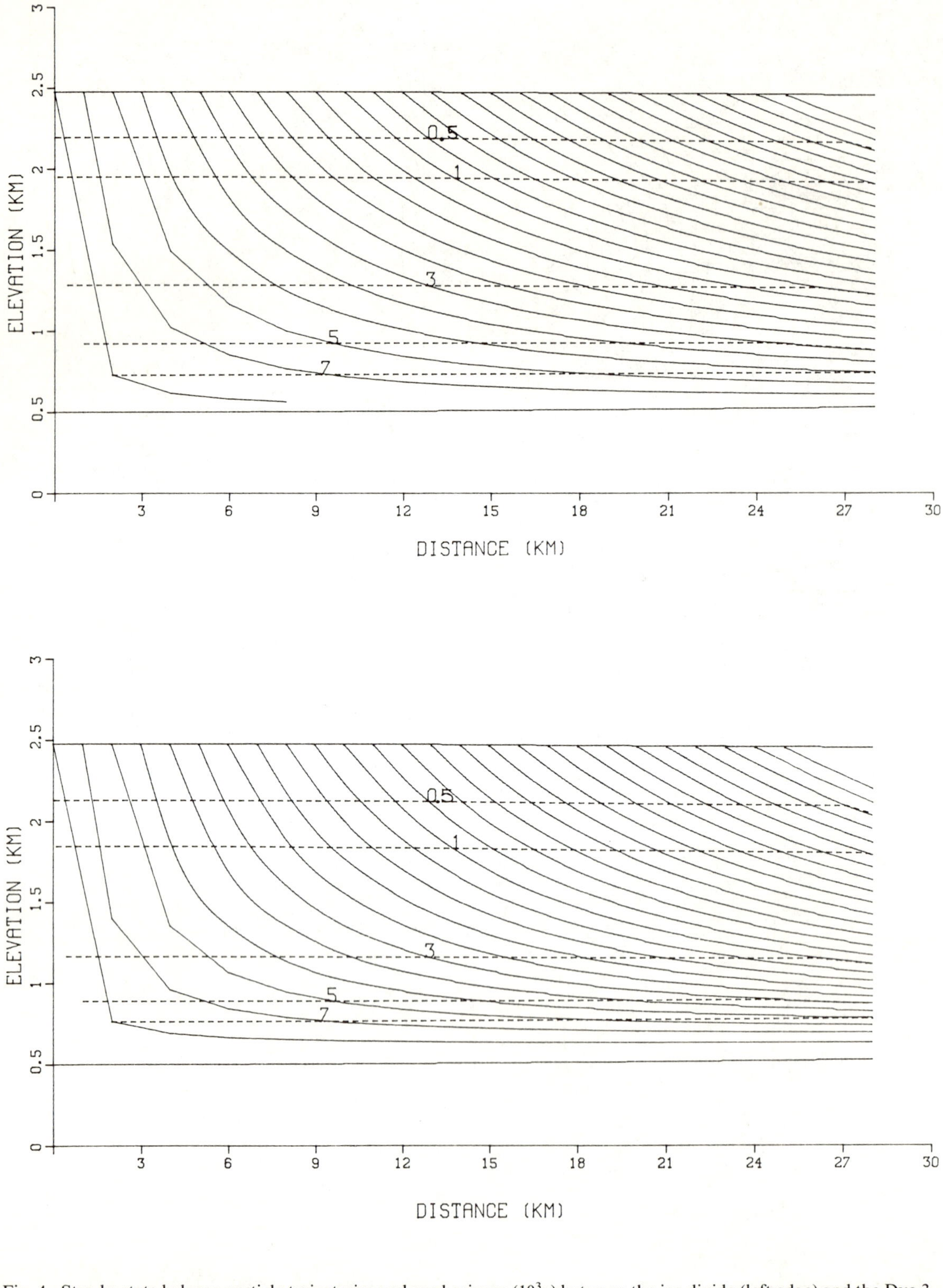

Fig. 4. Steady-state balance particle trajectories and age horizons (10^3a) between the ice divide (left edge) and the Dye 3 bore hole (right edge). The top diagram has been calculated with equation (5) for the vertical velocity w, and the bottom diagram with equation (6) for w'.

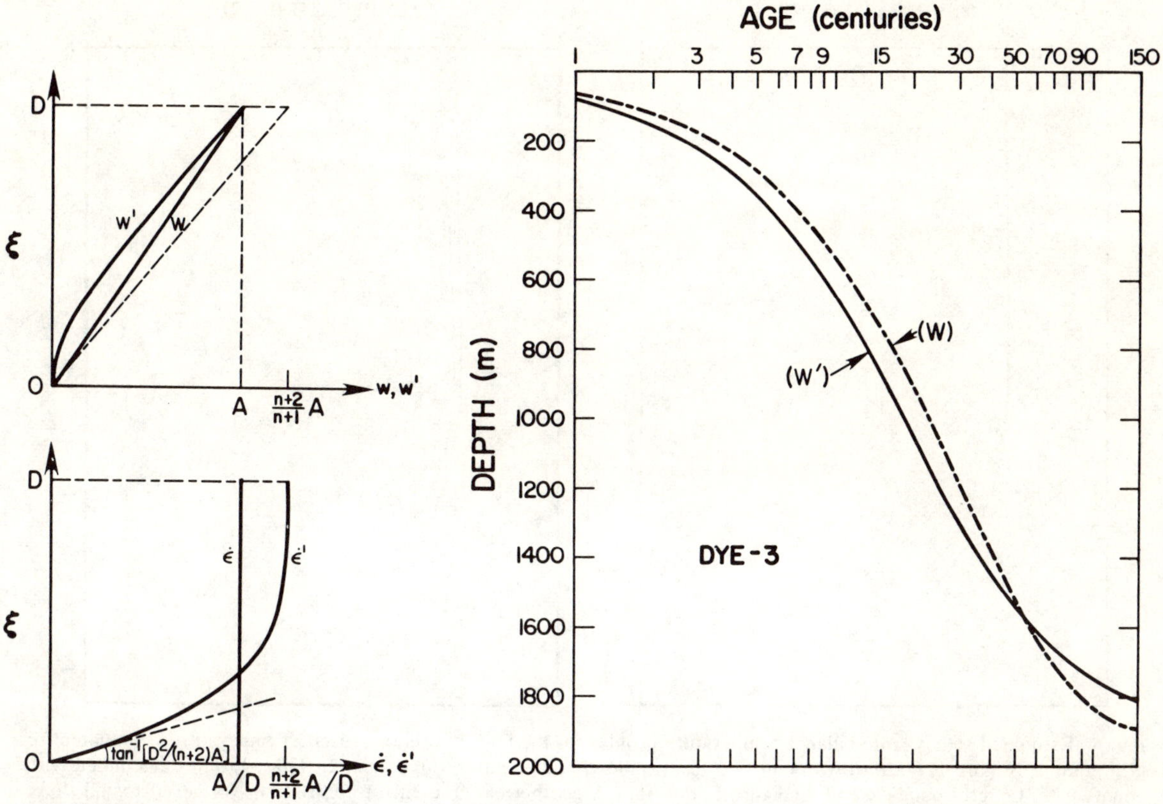

Fig. 5. Left: Vertical velocity profiles w (equation (5)) and w' (equation (6)), schematic. $\xi = D - z$ is the elevation above bedrock, $\dot{\epsilon}$ and $\dot{\epsilon}'$ are the vertical strain rates corresponding to the vertical velocity profile w and w', respectively. The broken line from the origin to the point $(n + 2)A/(n + 1)$ at $\xi = 1$ defines the surface gradient of the w' profile. Right: Steady-state balance age-depth scales for the Dye 3 borehole.

The lower boundary condition is derived from the basal heat input

$$q = \Lambda + \tau_b \bar{V}/JK \qquad (4)$$

where

Λ is the geothermal heat flux,
τ_b is the basal shear stress,
J is the mechanical equivalent of heat.

The use of the average velocity $\bar{V}$ in this context is justified since most of the shear is concentrated in the basal layer. The vertical ice velocity as a function of depth has been assumed in two forms:

(1) Linear

$$w = A\frac{D - z}{D} \qquad (5)$$

where

A is the mass balance,
D is the total ice thickness, and
z is the depth below the ice surface.

(2) Modified linear

$$w' = A\frac{D - z}{D}\cdot\frac{n + 2}{n + 1} - \frac{A}{n + 1}\left[1 - \left[\frac{z}{D}\right]^{n+2}\right] \qquad (6)$$

where n (= 3) is the exponent in Glen's [1955] flow law of ice.

The first of these vertical velocity profiles corresponds to a constant vertical strain rate.

$$\dot{\epsilon} = \frac{\partial w}{\partial z} = -\frac{A}{D}$$

The second form assumes, following Budd et al. [1976], that the vertical velocity decreases with increasing depth as does the average horizontal velocity due to shear deformation:

$$w' = A\frac{D - z}{D}\cdot\frac{\bar{V}_z}{\bar{V}_0} \qquad (5a)$$

where the overbars denote averages over the distances from the base of the ice to the levels D–z and the surface, respectively. This amounts to making the vertical flux w' through a unit area proportional to the horizontal flux below that level. Neglecting the small horizontal divergence of the ice flow around Dye 3 the velocity at depth z is found from the integral of Glen's law, viz,

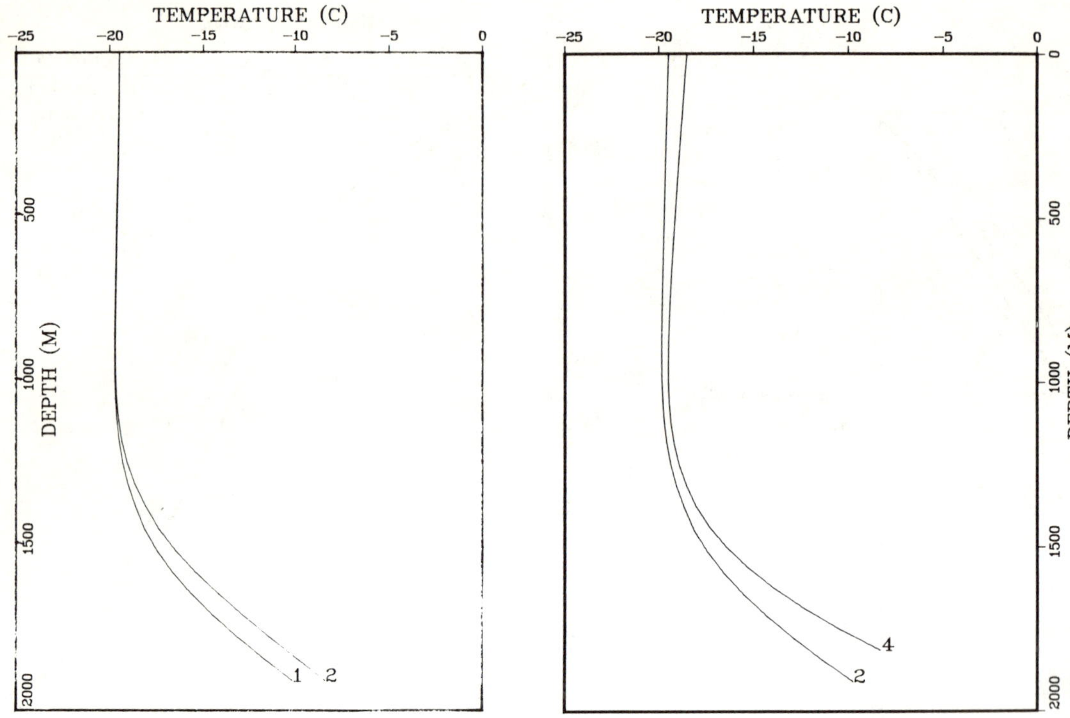

Fig. 6. Estimated steady-state balance temperature profiles for the Dye 3 boreholes. Left: Temperatures calculated for grid point 2, (Figure 1, bottom) with equation (5), profile 1, and equation (6), profile 2. Right: Profiles calculated with equation (5) for grid points 2 and 4 (Figure 1, bottom). A geothermal gradient of 2°C/100 km was used for all these calculations.

$$V_0 - V_z = B(T_1 \ldots)(\rho g \alpha)^n z^{n+1}/(n + 1) \tag{6}$$

where ρ, g, and α are the density, gravitational acceleration, and surface slope, respectively, and B is a function of temperature and other factors such as crystal fabric, impurities, etc. Figure 4, left, shows the two profiles of vertical velocity; the shape of the w' profile is very similar to that constructed for the Camp Century borehole from two linear strain rate profiles by Dansgaard and Johnson [1969]. Temperature profiles were computed with both w forms as a test of their sensitivity to the strain-rate parameter.

The vertical-velocity formulation also plays a crucial role in the construction of the ice-particle trajectories and in the derivation of age-depth scales for the ice core. The particle trajectories are composed of small segments, $\delta\chi = V_z \delta t$ (horizontal) and $\delta z = w\delta t = (w/V_z)\delta\chi$ (vertical). The age at the level z in the core results from integrating $\delta t = \delta z/w$ over the depth range from the surface at the particle origin to depth z in the borehole.

Results

The balance particle paths from the ice-sheet surface to the core site are shown together with age horizons in Figure 5 for the two types of vertical velocity profiles. The difference between the two age-depth scales emerges more clearly from the right hand side of Figure 4, which shows that equation (6) gives lower ages than equation (5) for the upper portion of the core and much higher ages for the bottom section, where, however, the assumption of steady-state conditions becomes increasingly unrealistic.

The corresponding borehole temperature estimates are shown in Figure 6. The profile labeled 1 in the left hand side of the figure was obtained using equation (5), and profile 2 using equation (6); their differences reach almost 2°C at the base of the ice. The profiles on the right in Figure 6 were calculated with equation (5) and relate to the gridpoints just upstream and downstream of Dye 3, cf. Figure 1 bottom; they show differences of the same order of magnitude.

Figure 6 can be interpreted in terms of the sensitivity of the balance model to the vertical velocity assumption and to errors in the input data. The calculation for the two gridpoints in effect involves a slight shift of the four input patterns in relation to the Dye 3 borehole site. In a personal communication, Reeh has pointed out that the contours upstream of Dye 3 are now better known and suggest a flowline initially directed north and originating further south than that here used (Figure 1, bottom). Moreover he expressed doubts about two large mass balance values used in the construction of the isopleths of Figure 2, upper left. However it is well known that the surface-mass balance can undergo substantial changes with small-scale topographic features. The mass balance isopleths in Figure 2, like the other features described in this paper, have resulted from an objective interpolation and smoothing procedure and cannot be adjusted arbitrarily.

The effect of the suggested upward bias on the thermal calculation would be to lower the estimated basal temperatures below those corresponding to a steady state with the correct mass balance field. Agreement of the erroneous temperatures with those observed could then cover up a genuine climatic warming. This underlines the limitations imposed by the relatively coarse net used and by the smoothing needed to represent the glaciological regime of the area.

Conclusion

It must be emphasized once more that the work here reported did not aim at predicting precise local features but at estimating the order of magnitude that certain physical characteristics of the ice sheet might take if it were in steady state or undergoing very slow changes with time. Moreover, special weight was placed on testing the sensitivity of the steady-state results to changes in parameters and input data. Two such tests only have been discussed, but it is easy to see how they could be expanded to take in all relevant parameters. However the next step clearly would be to use the observed features of the core and additional observations from its surroundings as starting points, both for the balance analysis and simulation of the ice history with various dynamic models.

Acknowledgments. This work was supported by Grant #DPP 7906789 from the National Science Foundation and by a grant from the Environmental Research Laboratories of NOAA. D. Jenssen of the University of Melbourne contributed in a major way to the development of the computer programs used. All this help, and that of I. Butler and B. McDonald in producing the paper, is gratefully acknowledged.

References

Budd, W. F., D. Jenssen, and U. Radok, Derived Physical characteristics of the Antarctic ice sheet MkI, *Publication #18,* Meteorology Department, University of Melbourne, 1970. (*Publication* No. *120,* Australian National Antarctic Research Expeditions, 1971.)

Budd, W. F., N. W. Young, and C. R. Austin, Measured and computed temperature distributions in the Law Dome ice cap, Antarctica, *Journal of Glaciology 16*(74), 99–110, 1976.

Budd, W. F., T. H. Jacka, D. Jenssen, U. Radok, and N. W. Young, Derived physical characteristics of the Greenland ice sheet, Mk I. *Publication No. 23,* Meteorology Department, University of Melbourne, 1982.

Dansgaard, W. and S. J. Johnsen, A flow model and a time scale for the ice core from Camp Century, Greenland, *Journal of Glaciology 8*(53), 215–223, 1969.

Glen, J. W., The creep of polycrystalline ice, *Proceedings of the Royal Society, London, A 22B,* 519–538, 1955.

Jenssen, D. and U. Radok, On the joint interpretation of total gas contents and stable isotope ratios in ice cores, *Annals of Glaciology, 3,* 152–155, 1982.

Radok, U., R. G. Barry, D. Jenssen, R. A. Keen, G. N. Kiladis, and B. McInnes, Climatic and physical characteristics of the Greenland ice sheet, *CIRES/ERL Climate Program Report,* University of Colorado at Boulder, 1982.

Robin, G. de Q., Ice movement and temperature distribution in glaciers and ice sheets, *Journal of Glaciology 2*(18), 523–532, 1955.

EPILOGUE

"A Mightly Maze! But Not Without A Plan."
. . . Alexander Pope

The science plan for the International Greenland Ice Sheet Program detailing the scientific objectives, rationale and outline of the complex instrumentation, data analysis and logistical needs was published in July 1976, some six years after a group of scientists and engineers from the United States, Denmark and Switzerland who had been involved with the analysis of ice cores obtained at Camp Century, Greenland, in 1966, met to outline its basic contents. One of the fundamental goals of the Greenland Ice Sheet Program was to identify sites in Greenland where the recovery of a complete column of ice from surface to bedrock could be arranged so that the bottom of the ice core would have the maximum geological age. By this means it was planned to obtain a complete sample of ice varying in age from the present to as far back in time as exists in Greenland. Sites were selected, but budgetary and logistical constraints dictated that the first core be obtained at a site known as Dye-3. This is not the optimum drilling site, but it is a site of sufficient interest to justify its selection and the installation of drilling equipment, living quarters, etc. to begin. Thus, Dye-3 was chosen as the first site for drilling through the Greenland Ice Sheet under the GISP program.

As a result of exceptionally effective and productive planning and execution by an international group of scientists and engineers, and an effort that now can be described as brilliant, a complete ice core was obtained. The papers in this volume result from a coordinated plan of analyses of this ice core and provide a "shower" of knowledge and information, in addition to providing corroboration of a number of conclusions and hypotheses that resulted from the analysis of the Camp Century core and various surface samples taken earlier over the Greenland ice sheet. More surely will follow.

Here is an instance in which epilogue, in a very real sense, is prologue; more remains to be done before the Greenland Ice Sheet Program plan will have been accomplished. It is a time to reaffirm the GISP objectives and reset our course. One is reminded of the lines from Joaquin Miller:

> Behind him lay the gray Azores,
> Behind the gates of Hercules;
> Before him not the ghost of shores;
> Before him only shoreless seas.
>
> The good mate said: "Now must we pray,
> For lo! the very stars are gone.
> Brave Admiral, speak; what shall I say."
> "Why, say 'Sail on! Sail on! and on!"

The drilling techniques developed especially for obtaining the Dye-3 ice core have been proven and are available for use at a more scientifically important site. The analytical techniques and instrumentation required for complete analyses have been developed and assembled. The logistical infrastructure exists in part and the remainder required can be assembled to support an operation at a more distant site and at the higher elevation required. Interest and enthusiasm on the part of the scientific teams that have participated in this initial phase is high; public interest also is high. The information and knowledge that will continue to accumulate as the objectives of the Greenland Ice Sheet Program are pursued and attained is of sufficient value to sustain the enthusiasm of the scientists and engineers involved and, in potential value, is more than sufficient to justify the necessary commitment of public funds. Let us not stop now.

Duwayne M. Anderson
Texas A&M University